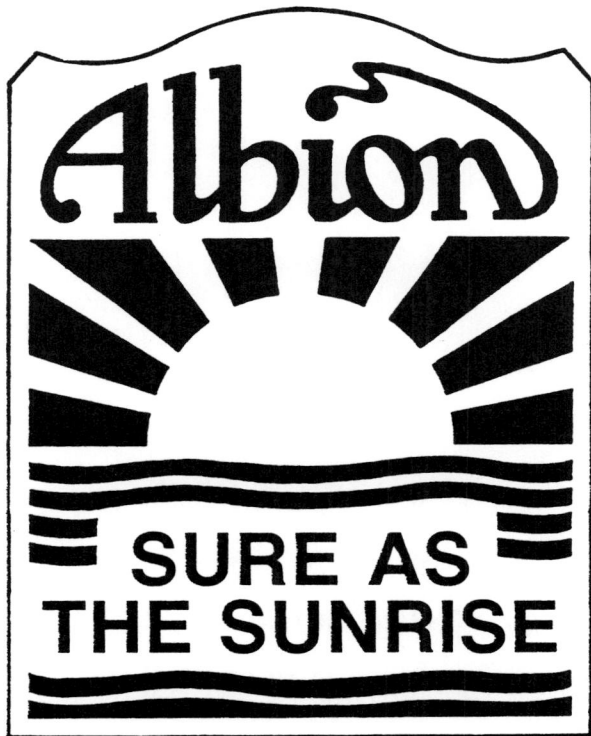

Albion

SURE AS
THE SUNRISE

Sure as the Sunrise

A History of Albion Motors

SAM McKINSTRY

JOHN DONALD PUBLISHERS LTD
EDINBURGH

ISBN 0 85976 473 7

British Library Cataloguing in Publication Data.
A catalogue record for this book is available
from the British Library.

Typesetting, graphics & PostScript origination by Brinnoven, Livingston.
Printed & bound in Great Britain by Hartnolls Ltd, Bodmin, Cornwall.

Preface

What is a company? A legal entity, certainly, a commodity that can be bought and sold and which generates returns for its owners, most emphatically. But a company is far more than these: it is a working environment with a character derived from its products, plant and buildings, a social experience, a collective memory, a reputation, a source of pride and self-fulfilment, even a symbol of national identity. In short, it is also an idea, a state of mind, partly subjective but to a large degree shared. Albion was and still is all of these, a representative of a dying breed both in Scotland and in Britain generally. Firms like it have been gradually replaced by plants arising from the 'lift and lay' policies of the multinationals on whom our industry now depends so much. These newer plants come and go in a period of time too brief for the loyalty, pride and sense of identity of the employees and the wider community to be shaped by their presence. The speed of technological advance and broader economic and managerial factors are of course to blame for the demise of the company-as-institution. The great shipbuilding firms of the Upper Clyde, premier examples of the genus, have all but gone, as have the steel mills and foundries that fed them. The glory and the grime have both departed. The chip, rather than the ship, rules the waves of Scottish manufacturing industry, leaving those involved in its production and use with a lot less to get excited about.

But today there still is, in Scotstoun, Glasgow a company which in spite of numerous legal and corporate reincarnations, retains the Albion identity and ethos in a line of succession unbroken for a century. Throughout the world, too, thousands of former employees, enthusiasts and owners still share the memories and lovingly restore and use Albion vehicles with undiminished affection.

The subject of their devotion is the subject of this book. It aims to capture some of the essence of Albion, including the visual and human dimensions, but in a way that does as much justice to the underlying management, technological and economic realities as the surviving records allow.

In a nutshell, while it is essentially a business history, the book seeks to satisfy all types of reader with an interest in the unique institution that is Albion. So as not to deter the non-specialist, references to documentary, bibliographical and other sources are dispensed with, but a full bibliography has been provided.

Glasgow 1997   *S. McK.*

Acknowledgements

My sincere thanks are due to a number of people who helped in many different ways with this book. Brian Lambie of the Biggar Museum Trust gave me freely of his deep and unique knowledge of Albion and allowed me unlimited (and even unreasonable!) access to the Albion Archives at Biggar, which also provided the bulk of the illustrations. In addition he read large tracts of the manuscript and made many useful suggestions.

Norman Lochhead of Leyland DAF Trucks shared mutual memories of Leyland Bathgate with me and provided copies of publications dealing with the reshaping of the old British Leyland Truck and Bus Division into the collection of companies that survives it today. Robert McCallum provided advice and contacts at Albion Automotive, and Scott Bissland, company secretary and financial controller there, shared Albion reminiscences and provided information, as did Elaine Catton-Quinn, marketing manager. Their cheerful help is greatly appreciated.

Special thanks are due to Pat Stevenson of Paisley University for converting my angular pencil drafts into beautiful wordprocessed pages with her customary skill and efficiency.

Many others helped in a host of ways too numerous to mention, and I thank them also for their assistance.

Contents

Preface v

Acknowledgements vi

Illustrations.. ix

CHAPTER ONE
New Century Dawning: Murray, Fulton and the Dog Carts 1

CHAPTER TWO
On the Road: From Cars to Commercials 16

CHAPTER THREE
Expansion 1910–1914: Van Makers to the Empire 32

CHAPTER FOUR
Albions at War: Production under Pressure 47

CHAPTER FIVE
Back to Peace: Boom, Bust and Buses 64

CHAPTER SIX
Radiations 90

CHAPTER SEVEN
The Turbulent Thirties 100

CHAPTER EIGHT
World War Two and After 123

CHAPTER NINE
The Leyland/Albion Era, 1951–1970 148

CHAPTER TEN
Sunset Meets Sunrise: Emasculation and Decline
under British Leyland 173

CHAPTER ELEVEN
After the Eclipse: The Rise of Albion Automotive 187

CHAPTER TWELVE
Inextinguishable Rays: Albion and the Biggar Connection ... 191

Bibliography 197

Appendix 1: An Albion Chronology 201

Appendix 2: Financial Details 207

Appendix 3: Output Statistics 213

Appendix 4: Order Book to July 1904 215

Appendix 5: Manpower Statistics 219

Index 221

Illustrations

1.1   Schedule for gasfitting at Albion's Finnieston works, 1900.

1.2   Work-in-progress at Finnieston, 1900.

1.3   Dog cart chassis, showing the robust construction.

1.4   Receipt from John L. Murray for deposit on his dog cart of £100 (1900).

1.5   Albion's first publicity brochure, showing Fulton and Murray on a dog cart, 1901.

1.6   Albion's Finnieston workforce in 1903 (D. Keachie in bowler hat), with car lift visible.

1.7   Tonneau version of dog cart outside Finnieston c 1902.

1.8   Dog cart 2 cylinder horizontally opposed engine.

1.9   John Lawson, chauffeur to John L. Murray, Biggar at the tiller of a dog cart, with N. O. Fulton alongside, 1900.

1.10 John Lawson and Mary G. Murray (front car) with Mrs T. B. Murray and Mrs N. O. Fulton (rear car) at Heavyside, with John L. Murray standing.

2.1   Albion's new premises at Scotstoun, 1903.

2.2   Cancelled Albion share certificate, 1903.

2.3   10 h.p. tonneau at Scotstoun, 1903.

2.4   A dog cart in use at Biggar. W. Ovens and boys collect for a bonfire c 1903.

2.5   John Francis Henderson and wife in an A6 Albion at Cairn-o-mount trials, c 1906.

2.6   John Buchan and wife in rear of dog cart at the Peebles electoral campaign, 1903.

2.7   16 h.p. Albion van c 1905.

2.8   Albion/Lacre van in Harrods livery c 1908.

2.9   John Lawson at Heavyside with Albion 16 h.p. tonneau, 1906.

2.10 Sir T. D. and Lady Gibson-Carmichael at Melbourne in Governor's Albion Landaulette, 1910.

3.1   15 h.p. 15 cwt. van in livery of Ardath Tobacco Company Co. London, 1911.

3.2   The precursor — Albion's first export, sent to Kuala Lumpur in 1901.

3.3   Albion offices and factory as constructed by 1913.

3.4   The new Albion office block in 1913.

3.5   Albions in livery of John Barker & Co. c 1913.

3.6   Part of Harrods' fleet of 70 Albions, 1913.

3.7   Page from Albion's 1913 sales brochure.

4.1   Chassis assembly, 1914.

4.2   Women shell production workers, 1916.

4.3   Line up of Albion War Office vehicles, 1915.

4.4   15 h.p. Albion ambulance, 1916.

4.5   A wartime Albion A10.

4.6   Wartime convoy en route via Biggar in 1915.

4.7   Biggar residents inspect War Office vehicles, 1915.

4.8   Bassett Lowke model of Albion A10 war vehicle.

4.9   T. B. Murray in DSc academic dress, 1917.

4.10 Albion A10 mobile workshop.

4.11 An A10 in Ankara, recently photographed — a survivor of Gallipoli?

4.12 Albion A10 line up at Scotstoun.

5.1   An A10 Albion leads the first post-war Glasgow Fair Celebration, Stranraer, on 19 July 1919.

5.2   Thomas Hamilton of Fauldhouse's 'demobbed' A10 military waggon, in use for haulage c 1920.

5.3   1923 24 h.p. military waggon.

5.4   T. Blackwood Murray, late 1920s.

5.5   A10 lorry supplied to South African Railways in 1918 (photographed in 1932).

5.6   War Office subsidy chassis outside Alexander McAra's garage, Dundee c 1923.

5.7   General view of McAra's, Dundee with Albions outside, c 1923.

5.8   Two Albions, one Maudslay and one Ford van at Rowe Brothers, Bristol, 1920s.

5.9   1923 Albion photographed at Bombay docks in 1958.

5.10 18 passenger Viking coach, 1923.

5.11 First non-stop London to Glasgow bus, Model 26, 1925.

5.12 Albion Model 27, 3 ton tanker in livery of Shell, Melbourne, 1926.

5.13 B.P. overtype tanker, late 1920s.

5.14 Carter Paterson overtype truck, late 1920s.

5.15 Albion 20 seater bus at Scotstoun, late 1920s.

5.16 Albion bus in Redland livery, late 1920s.

5.17 20 h.p. tractor and trailer, 1920s.

5.18 Albion Model 40 bus owned by City of Oxford Motor Services, 1929.

5.19 Promotional drawing of a Model 28 Viking Six in City of Oxford livery, 1928.

5.20 20/36 h.p. school ambulance for Dumbarton Education Authority, 1929.

5.21 Lorry in Scottish Co-op livery, 1920.

5.22 The Prince of Wales and N. O. Fulton at Scotstoun.

5.23 The Prince of Wales and N. O. Fulton with David Keachie at far right.

5.24 Fulton and the Prince of Wales examine machinery.

5.25 The Prince of Wales examines a dog cart and an A10 war waggon.

6.1 Layout sketch for Albion recreation ground, Yoker.

6.2 Bowling Club Committee and Recreation Association officials (including Keachie and Mr & Mrs N. O. Fulton), 1928.

6.3 Motor cycle and car club at Rest-and-be-Thankful (from *Radiator*).

6.4 'Team Work' (*Radiator*).

6.5 'The Watta-Lily', 1929 (from *Radiator*).

7.1 An Albion on the Glasgow/Walkerburn route, c 1930.

7.2 Albion trailer set, c 1930.

7.3 John Francis Henderson (from a 1924 portrait).

7.4 Albion rigid six wheeler of R.A.S.C. Motor Transport Department at Hounslow Heath.

7.5 Albion charabanc at Lincoln's Inn Fields, London, early 1930s.

7.6 33 seater Albion in livery of Brighton Safety Coaches c 1934.

7.7 Model 69 Valkyrie in Strachan's of Deeside livery, early 1930s.

7.8 Valiant Model M70 32 seater bus supplied to Red and White Services, Lydney, c 1934.

7.9 Valkyrie 32 seater in livery of Dodds of Troon, September 1934.

7.10 A 6 W Valkyrie in Young's livery at work in Paisley, 1933.

7.11 A Viking Six in Young's livery leaving Clyde St., Glasgow for Largs, October 1931.

7.12 Albion bus on tilt test, c 1935.

7.13 N. O. Fulton.

7.14 Albion bus in Midland livery, mid 1930s.

7.15 Albion stand at the Rand Motor Show, April 1936.

7.16 Albion Model 34 at work in South Africa, late 1930s.

7.17 Chassis department, Scotstoun, 1934.

7.18 Machine department, top flat, Scotstoun, 1934.

7.19 Bob Coutts delivering a chassis to bodybuilders in Portsmouth, 1937.

7.20 H. E. Fulton.

7.21 M127 $5\frac{1}{2}$ ton lorry, built 1935-41.

7.22 Model T561 14 ton Gardner engined van for Melrose's Tea, 1937.

7.23 Model T561 with Gardner engine, 1938.

8.1 BY3 truck for the Ministry of Supply, 1940–41.

8.2 Ambulance AM463 as used by the Air Ministry.

8.3 BY5 lorry with folding boat equipment, 1941–45.

8.4 CX33 8 wheel drive tractor.

8.5 Standard AM463 Air Ministry tanker.

8.6 Albion BY1s at Loughborough, 1939.

8.7 AM463 articulated lorry supplied to the Air Ministry.

8.8 FT11 4 WD truck for the Ministry of Supply.

8.9 WD CX 24 S tank transporter.

8.10 BY1 petrol tanker for the Air Ministry.

8.11 6 wheel Albion WD truck.

8.12 Exhortation to boot Churchill out in *Sunrise,* 1945.

8.13 Visit of concessionnaires and official repairers to Albion, September 1948.

8.14 6 passenger station waggon, 1948.

8.15 Brochure for the Clydesdale tractor, FT 101 and 102, c 1950.

8.16 A SCWS Albion c 1950.

8.17 An Albion demonstration truck in South Africa, c 1948.

9.1 Truck assembly at Scotstoun, 1955.

9.2 Albion at the Scottish Motor Show, 1957.

9.3 Albion Claymore with box van body, mid 1950s.

9.4 A Caledonian on the Albion stand at the 1957 Scottish Motor Show.

9.5 H. W. Fulton and J. L. Murray on an A6 Albion outside the Kelvin Hall, 1957.

9.6   FT 39 Albion Victor bodied as a pantechnicon at Brisbane, late 1950s.

9.7   Albion Victor VT 17N supplied to Ceylon Transport Board, 1959.

9.8   Albion Victor VT 17 N for Boston Transport, Barbados, 1959.

9.9   Albion Caledonian tanker, 1960.

9.10 LAD cabbed Chieftain in 1958.

9.11 Unidentified Albion chassis at Scotstoun, 1958.

9.12 Double reduction rear axle, as fitted to the Chieftain and Victor range from 1960.

9.13 9 speed gearbox Type GB 248, optional on Clydesdale and Reiver models from 1960s.

9.14 Super Clydesdale diesel tractor with Ergomatic tilt cab, late 1960s.

9.15 West Gate, Scotstoun works, late 1960s.

10.1 Signs of the times — Albion exhibits as part of B.L. at the 1969 Scottish Motor Show.

12.1 T. B. Murray and his new wife in Paris in 1900.

12.2 Invitation to the T. B. Murray centenary exhibition, Biggar, 1971.

12.3 J. B. Murray, R. C. Dougal and L. Capaldi at the Murray Centenary exhibition, 1971.

12.4 The Scotstoun works date stone re-installed at Biggar, 1989, with, left to right, Brian Lambie, Lady Cowie (granddaughter of T. B. Murray), Grizel Hoyle and Jim McGroarty.

CHAPTER ONE

New Century Dawning: Murray, Fulton and the Dog Carts

The log book kept by Albion Motors until the 1970s records that it began its existence as the Albion Motor Car Company on 30 December 1899. Its inception cannot in fact be dated with this level of precision. The deed of partnership which formally united the two participants, Thomas Blackwood Murray and Norman Osborne Fulton, was dated 8 February 1900. It implies that the partnership commenced on the first day of the year 1900. A further complication arises from the fact that a letter written by Murray on 6 December 1899 in connection with the purchase of new machine tools was sent in the company's name. What is plain is that the launch of the new business was the culmination of several years of planning before the close of the old century. It is also clear that it was highly appropriate to choose the dawning of the new century to inaugurate officially a venture that was based on a revolutionary technological advance set to change the face of the civilised world — the advent of the motor vehicle.

Who were these men? Murray, born in 1871, was the son of the local architect in Biggar, Lanarkshire, John Lamb Murray, whose design skills can still be seen in the lofty gothic Gillespie church whose broach spire still dominates the main street of this beautiful, unspoilt country town. In 1869–70 he had acted as clerk of works to the leading Scottish architect of the period, David Bryce, who was carrying out alterations at the ancient Biggar Kirk. Murray senior subsequently went on to design schools, churches and large and small houses throughout Lanarkshire. One of his largest commissions was for the new Hartwood Hospital, and it was he who was responsible for the Biggar water supply, which consisted of a seven-mile pipeline and storage tanks, still giving satisfactory service today. Murray's seemingly boundless versatility also extended to farming Heavyside, the family home, and in addition he was estate factor to the Skirling, Castlecraig and Carmichael estates. These activities must have made John Lamb Murray a man of some means.

Young Thomas was educated at Biggar School, and then at George Watson's, Edinburgh, a famous fee paying institution, and from there took a BSc degree at Edinburgh University which he completed in 1890, having specialised in the exciting new discipline of electrical engineering. That same year he had perfected and patented a water turbine governor for use in hydro-electric generating plants.

His first post was with King, Brown and Company of Edinburgh and a second post with Rankin Kennedy of Glasgow, as head draughtsman and assistant, was held until about 1892. This work involved the design of dynamos and electric motors and production management responsibilities for these machines. Afterwards, he completed consultancy work for Edinburgh corporation, advising it on the use of waste heat for driving electrical machinery at Powderhall Refuse Disposal works, and in connection with lighting and tramway stations. Between 1893 and 1896 he worked with Mavor and Coulson in Glasgow as Manager of the Installation Department. In this post he was involved in producing an electrical ignition system for the new Coventry Daimler motor car company, and while employed with Mavor and Coulson, he had invented the low-tension magneto, which was a more efficient alternative to the unreliable incandescent tube or accumulator ignition systems universally used on the first generation of motor cars. In Murray's device, rotating magnets carried on the crankshaft induced currents in stationary coils. The system was so successful that it was to persist in Albion vehicles for many years after the development of the modern high-speed magneto. By the middle of the decade, this brilliant young man's turbine governors were in place as far afield as Tasmania, and via Murray senior's architectural practice, could also be found at the nearby Douglas Castle, a country house belonging to the Earl of Home. Clearly the practical, theoretical and managerial talents of Murray senior had been handed on and in some areas considerably amplified in his son.

Norman Osborne Fulton had progressed by a different and less spectacular route. He was the son of a wholesale provision merchant in Glasgow and had been educated at the prestigious Allan Glen's School. He had served an engineering apprenticeship with Kesson and Campbell of Shettleston, manufacturers of pithead machinery, and had studied in evening classes at the Andersonian College, now Strathclyde University, eventually becoming a member of the Institution of Mechanical Engineers. Possibly drawing on parental connections, Fulton gained his post-apprenticeship experience in food processing businesses, which included a creamery. Thereafter he enhanced his production engineering and managerial experience in the chemical industry. Fulton's interests were different from Murray's, but very compatible. Well organised, rational and efficient production systems concerned him more than innovations in product design.

Fulton had in 1896 set out on the path that would eventually bring him into contact with Murray. That year he joined his cousin, George Johnston at the Mo-Car Syndicate in Camlachie, Glasgow. Johnston is said to have built a motor car which predated the vehicle built by Lanchester in England that is assumed to have been the first British-built automobile. Prior to this, he had successfully designed self-propelled steam trams, but his plans were discarded by his sponsors, Glasgow Corporation, in favour of electric trams operating from overhead cables. The Mo-Car Syndicate had been formed in 1895 to develop and produce Johnston's motor cars and its funding came from two main sources, Sir William Arrol the successful engineer, and the Coats family of Paisley, proprietors and founders of J. and P. Coats the huge multinational producers of sewing thread.

By November 1896 Murray had been hired by Johnston to take charge of the 'electrical department'. Johnston was Mo-Car's Managing Director, and he had given him a three-year contract, at an annual salary of £300. Fulton was already in position as works manager. Murray and Fulton seemed to hit it off immediately, for by the middle of June 1897 they were both holidaying in the Black Forest, on bicycles. Such pursuits required more than a passing fondness for this mode of mass transport, which had only a few years earlier emancipated those unable to afford horse-drawn conveyances. Murray had himself designed a form of caliper brake for the bicycle in 1890 that was suited to the newly introduced pneumatic tyre.

Once the initial euphoria of working together at Mo-Car was over, Murray and Fulton began to be disillusioned at the speed of progress there. An electric vehicle, on which Murray had been working, was abandoned because of the limitations of battery power. Murray, at least, was of the opinion that Johnston was too slow in tackling technical problems relating to the new car they were to produce. Out of sheer frustration, he had secretly applied for another post, assuring his prospective employers somewhat prophetically that if appointed he would make 'both an electrical and commercial success' of their business. The two friends instead made a pact to the effect that Murray would see out his contract, while Fulton would spend a year in the United States studying automobile manufacturing technique, which had already leapt ahead of Europe in its ability to produce large batches of vehicles. Certain key firms were making breakthroughs in manufacturing engineering practice. One of the most important in this respect was the Pope Manufacturing Company of Hartford, Connecticut, which had originally been a bicycle manufacturing concern. It had moved on to both electric and petrol-driven cars by 1896, and these sold in large numbers under a variety of names. It was to this business that Fulton was attracted, taking up a position as a mechanic in 1898.

Manufacturing methods at Pope had developed along the lines of engineering practice in the US armory. This involved specially made machine tools and fully interchangeable parts produced with gauges and to specified dimensional tolerances. The new approach contrasted with general engineering practice in Europe, in which components were not exactly matched, leading both to manufacturing problems and problems with the supply of spares, which did not always fit. The innovations mentioned were therefore indispensable to the development of large-scale production, a fact both Fulton and Murray came to understand very well.

How Johnston felt or what he knew about the reasons for the defection of Fulton to the United States remains a mystery. He is likely to have known that an independent motor car producing business lay at the back of his cousin's mind. If he did, his family relationship and his kind disposition would have made it difficult for him to do anything about it, even if that were possible.

Between the latter part of 1898 and 1899, Fulton and Murray were working on designs for their own car, which was already being slowly built. A letter from Murray to Fulton of 15 February 1899 reveals that the crankshaft for its petrol engine was being turned in John L. Murray's hard pressed Biggar workshop,

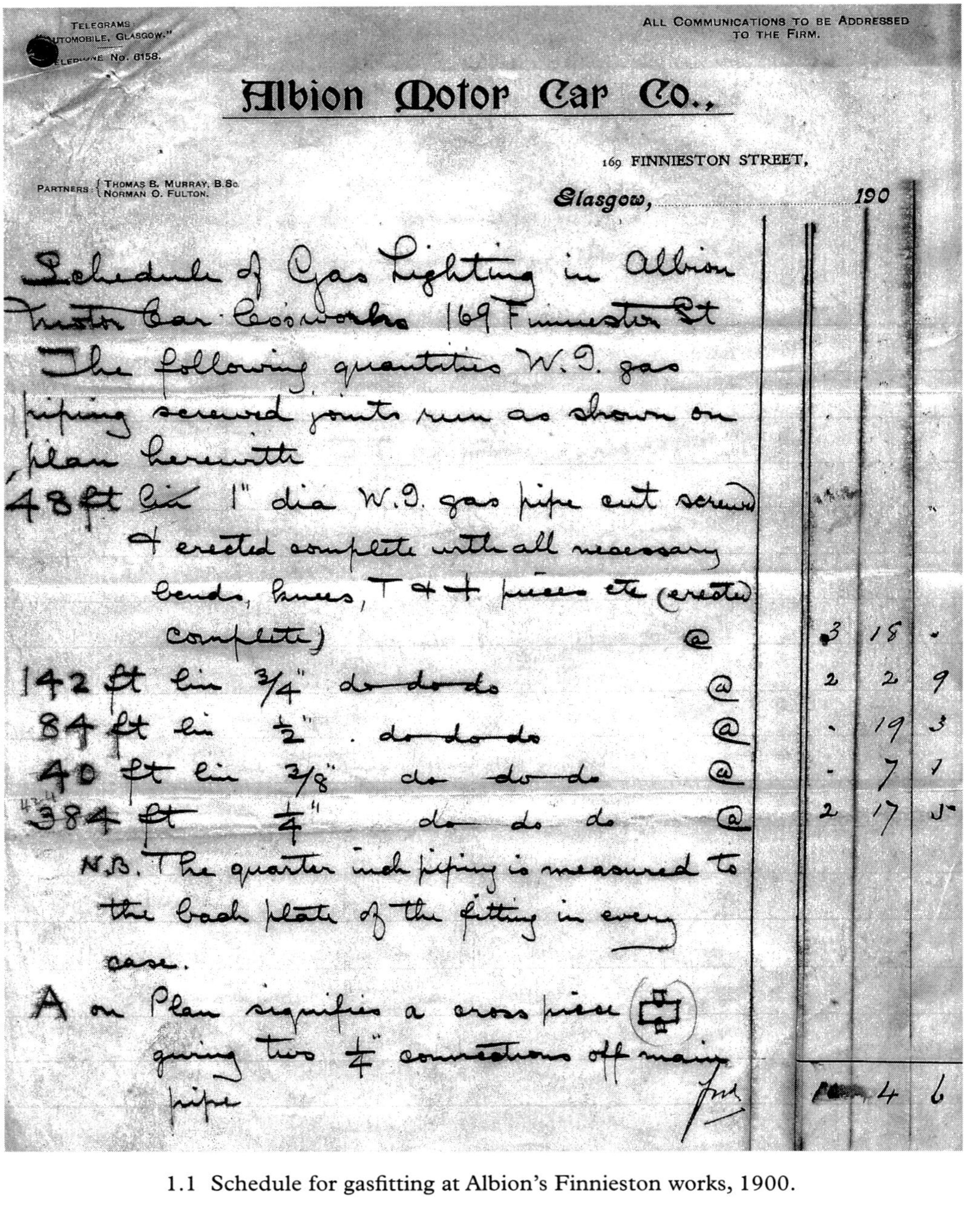

Albion Motor Car Co.,

169 FINNIESTON STREET,

PARTNERS: { THOMAS B. MURRAY, B.Sc.
{ NORMAN O. FULTON.

Glasgow, ... 190

Schedule of Gas Lighting in Albion
Motor Car Co works 169 Finnieston St
The following quantities W.I. gas
piping screwed joints run as shown on
plan herewith

48 ft lin 1" dia W.I. gas pipe cut screwed
& erected complete with all necessary
bends, knees, T & + pieces etc (erected
complete) @ 3 18 -

142 ft lin 3/4" do do do @ 2 2 9

84 ft lin 1/2" . do do do @ . 19 3

40 ft lin 3/8" do do do @ . 7 1

384 ft 1/4" do do do @ 2 17 5

N.B. The quarter inch piping is measured to
the back plate of the fitting in every
case.

A on Plan signifies a cross piece
giving two 1/4" connections off main
pipe

4 6

1.1 Schedule for gasfitting at Albion's Finnieston works, 1900.

1.2 Work-in-progress at Finnieston, 1900.

where gears were also being cut and con rods were being made. Letters during the summer record a request from Murray that Fulton ascertain how carbon be cleaned out of cylinders, and queries on engine and steering details. For his part, Fulton was sending across sketches of axle components. In September 1899, Murray was notifying Fulton of problems with engine castings, which were causing difficulties with the machining of cylinders, but expressed the hope that the first engine would be running by the end of the year. As Murray sent over progress reports from Biggar, he clarified the basic strategy for the launch of the new company: to produce a quiet vehicle, the first unit of which was to be built 'as cheaply and as rapidly as possible'.

Fulton's practical abilities were underscored by the fact that on announcing his intention to return to Scotland, he had been offered an increase in salary at the Pope Company, as an incentive to stay. Unknown to his new employers, he had been secretly shipping sample wheels and axles to Murray. On 18 November 1899, Fulton left New York for Scotland, and a few months later, the partnership was formally constituted.

The new business took up residence at 169 Finnieston Street, Glasgow, on the edge of the docklands of the upper Clyde and in the midst of all manner of engineering works. The premises it occupied had formerly been the repair shop of the Clan Line, a steamship company. Very soon, in 1901, Halley, a manufacturer of steam lorries, would take up residence in a different part of the same building. At first there were only four machine tools, a medium lathe, a heavy lathe, and

1.3 Dog cart chassis, showing the robust construction.

two drill presses. Here the seven employees, whose first weekly wage bill totalled £9-10-3, began to make the earliest vehicles.

These were of 'dog cart' design. The dog cart was a light but robust horse-drawn vehicle that had been popular throughout the 19th century for shooting expeditions and which provided back-to-back accommodation for driver and passengers. While some of the small Scottish motor car companies that were about to spring up close by would also choose this style of bodywork, it must be recognised that, elsewhere, it was being superseded by less derivative body types. Its selection for Albion cars reflects several local factors: it was chosen with an eye to the country house market, to which it would undoubtedly appeal, and its tough construction could take the battering meted out by the rough roads which still characterised much of Scotland. More to the point, perhaps, the Mo-Car vehicle on which the partners had worked, and which they knew best, was also designed as a dog cart. The engine was located in the compartment between the rear seats of the vehicle, formerly reserved for the dogs. In common with the Mo-Car vehicle, this was a horizontally opposed piston type, of two water-cooled cylinders, which was quite frequently found among the first generation of motor vehicles. It generated 8 h.p. Naturally the car was started by means of a Murray low tension magneto, which was activated either by a starting handle, or by pulling a rope with a wooden grip located at the driver's feet. The car was also provided with a Murray engine speed governor and an Albion-Murray lubricator, which, operated by a cam, force fed oil into the engine to prevent seizure. While the engine (and for that matter, most of the rest of the car) had quite definitely not been designed from first principles, the latter devices, patented by Murray, were

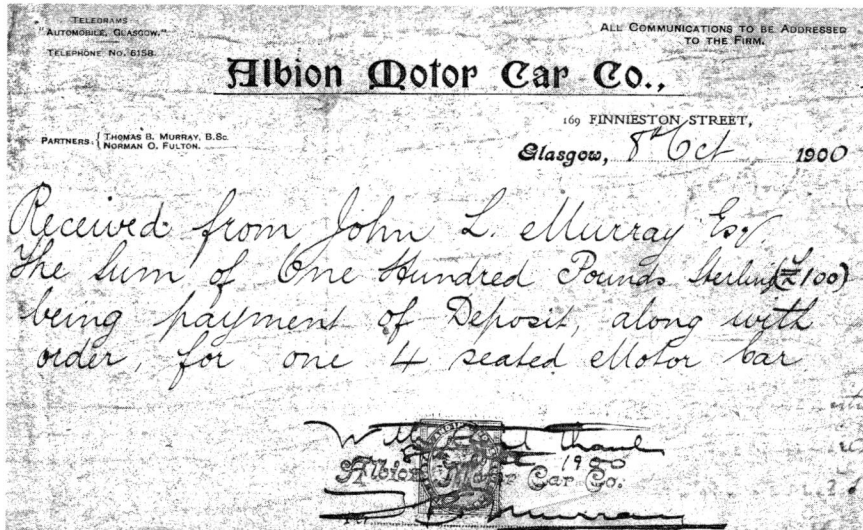

Receipt from John L. Murray for deposit on his dog cart of £100 (1900).

important steps in the direction of reliability, and would in due course pay lasting dividends in terms of the company's reputation.

The car was chain driven, and had two forward gears, capable of speeds of about 5 and 13 m.p.h., and one reverse gear, operating at 3 m.p.h. This tough ensemble was mounted on a robust frame of steel channel and angle iron. A 'Time book' from 1900 indicates that the work force consisted of fitters, turners, smiths, labourers and strikers. The first two trades must have produced engines and gearboxes, the latter three the chassis. In the beginning it is likely that there was only one representative of each.

The first order was placed by John L. Murray himself, a wise decision for purposes of customer feedback. A second-hand account of the tribulations associated with the start-up of the first engine at Finnieston, no doubt burnished and embellished by the passage of time, gives something of the flavour of the moment. James Livingston, writing of his father Hugh's involvement in this event, recalls that he was summoned from his nearby employment to the Albion workshop, where Murray and others were working on what appeared to be a 'block of metal anchored to its housing'. Beside this lay a pile of broken pipes. A new pipe was screwed into the block and a rope was attached to its end. After a cry of '1, 2, 3, go!' from Murray, the engine exploded into life, fracturing the new pipe. On examining these pipes, Hugh Livingston, who had noted that they had been made by 'amateurs', undertook to make a replacement in heavy copper, 'with flanges properly brazed'. In due course this was fitted to the 'block of iron' and after a further '1, 2, 3, go!' the engine exploded into life and began to purr sweetly. At this point Murray and his companions are said to have gripped hands and danced round both the 'purring iron' and Hugh Livingston, whose reward was a new job with Albion.

1.5 Albion's first publicity brochure, showing Fulton and Murray on a dog cart, 1901.

The only other glimpse of production in the Finnieston premises that is now available comes from an old photograph, taken from the firm's first publicity brochure. Here the neat line of chassis attests to the organised production philosophy of Fulton, confirmed on his sojourn in the U.S.A.

While John A. Murray placed the first order for one of these cars on 4 January 1900, he did not receive it until 1 February 1901. Clearly most of the first year of the operation was a period of learning and experimentation. By the autumn of that year, however, the company was sufficiently in command to take orders from the public at large. W. A. Verèl, of Newlands, Glasgow, placed his order on 12 September 1900 and took delivery of his vehicle on 29 December. Walter Graham of Bearsden placed an order on 10 September and received the vehicle on 26 January 1901, four days after the death of Queen Victoria. Murray senior took delivery of his vehicle on 1 February 1901. In all, six orders were taken in 1900, five of these being delivered in the spring of 1901.

In 1949, Jackson Millar, who was then chairman of Albion Motors, recounted that his father had been one of the first purchasers of an Albion vehicle. The records reveal that he ordered it on 5 February 1901, and had it delivered on 7 March 1901. He found that while the car ran well, it had a tendency to 'buck like a bronco' when the cone clutch was engaged. This was not untypical of cars of this era, and it did not deter Millar from injecting capital into the company shortly thereafter. It is not known whether Millar had any social connection with Fulton or Murray, but this factor, plus an element of personal recommendation seems to be responsible for some of the sales made in the company's first few years of operation. Many of the customers were from the prosperous suburbs of Glasgow. An increasing number would be attracted by the firm's early publicity brochures, the first of which seems to have been produced towards the latter part of 1901. The front cover shows Fulton (nearest the camera) and Murray at the tiller of a dog cart. Free trials were available by appointment, and correspondence was 'solicited'. Inside the brochure Albion's 'HIGH CLASS MOTOR CARS' were introduced to readers. They had been 'specially designed for the use of Medical and Business Men', and could be fitted with hoods when required. Particular attention was drawn to the fact that the cars were not 'voiturettes', an early form of vehicle which provided an engine and four wheels on what was essentially a bicycle-style frame, and which lacked robustness. Full opportunity was taken to explain that the 'Latest Up-to-date Machine Tools' enabled Albion to turn out 'the very highest class of work', and that 'all parts are made to gauges and are therefore interchangeable'.

This extremely solid motor car was offered at a basic price of £280 in dog cart form, and for £10 extra, could be had in 'tonneau' form. Extras available included Morocco leather upholstery (£5), an accelerator (£6) and a starting mechanism operated from the driver's seat (£6). A 'doctor's car' version was offered at £274 basic, and consisted of a dog cart minus the rear passenger bench. A 'Victoria Hood of best leather', resembling a large perambulator hood, could be added for £20. Finishes available were either natural varnished wood or paintwork to 'first class carriage' standard.

1.6 Albion's Finnieston workforce in 1903 (D. Keachie in bowler hat), with car lift visible.

The availability of tonneau bodies raises an important point. By the end of 1903, out of 112 vehicles sold, some 60 were tonneaux and 29 were dog carts, with the balance either sold as chassis for custom bodying or as vans, broughams or wagonettes. A crude sketch of a tonneau outline survives in the Albion archives and makes it clear that in this format, the vehicle had two forward facing rear seats but was otherwise similar to the dog cart. From the start, it was obvious that neither Fulton nor Murray with their engineering predilections had much of an interest in bodywork. This they left to Penman of Dumfries (who are still in business today) and Alex Cree of Glasgow appears also to have played a part at an early stage. The lack of surviving evidence of in-house bodybuilding activity indicates firstly, that these contractors shipped in finished bodies to Albion, who bolted them on to the completed chassis, and secondly, that detailed body design was carried out at the bodybuilders' offices.

If bodywork was not to the taste of Fulton and Murray, what was? Albion's founding partners were engineers whose preferences lay in the direction of simplicity, robustness and reliability (in Murray's case) and in productive efficiency (in the case of Fulton). In spite of Murray's excellent peripheral innovations in engine accessories, which stemmed from his fascination with rotary devices working off the back of prime movers, it is more accurate to describe him

10

1.7 Tonneau version of dog cart outside Finnieston c 1902.

as an improver of engines rather than an innovator in engine design. He was a follower rather than a leader, a key example of this being his use of a tiller on the new cars. Steering wheels were already beginning to predominate in other makers' designs. Even so, Albion praised the superiority of the tiller over the steering wheel in some of its early publicity literature. Soon it had to bow to market forces. Nevertheless, its conservative orientation began to pay dividends: as was noted earlier, it would produce more reliable vehicles. In due course, it would lead Albion away from motor cars.

By the time Albion had settled into steady production in the autumn of 1900, it had a handful of serious rivals in Scotland. Gottlieb Daimler's first successful engine of 1884 and Carl Benz' first motor car of 1884–5 had started an industry that by the turn of the century was flourishing on the continent, most notably in France, where Renault and de Dion had become leading makes. The European example had inspired English firms such as Daimler, Humber, Star, Sunbeam, Swift and Singer to enter the market, together with a whole host of less well-known names. One estimate suggests that in England about 50 firms had entered automobile production by this point, mostly on a tiny scale. Some of these firms had a background in cycle manufacture, some in other types of engineering, and many were clustered together in the Coventry/Birmingham area. In Scotland, Mo-Car was already active, as was the Hozier Engineering Company of Bridgeton, which was selling one or two voiturettes per week under the Argyll marque by

11

1.8 Dog cart 2 cylinder horizontally opposed engine.

1901. John Stirling of Hamilton had already adopted a different route into motor manufacturing, purchasing chassis from Daimler, Panhard and other continental sources and marrying these with coach-based bodies produced in his own coachworks, and in Edinburgh, the Madelvic Company had been selling an electric-powered Brougham carriage propelled by a central fifth wheel. A steam-powered car had even put in an appearance in Stirling, designed by John Simpson. In the fulness of time, events would prove that a wholehearted commitment to internal combustion design, wedded to substantial in-house manufacture, was necessary for long term success, as was a reducing dependence on French and German engineering, which was widely and quite blatantly copied or adapted in Britain. Another important ingredient for long-term success was also present at Albion: the production of standardised, interchangeable parts, modelled on U.S. practice.

At the beginning of the new century motoring was something of an adventure, reserved for the wealthy. Motor cars would, typically, cost from £150 for a light voiturette to £300/£400 for a substantial mid-range vehicle, equivalent to the annual salary of a successful 'professional man'. Up until 1896, it was illegal to drive a car unless preceded by a pedestrian carrying a red flag, and between 1896 and 1904, the speed limit was 12 m.p.h. Rudimentary suspensions, brakes and tyre treads combined with rough road surfaces to produce frequent accidents, and for this reason, the car was not popular with other road users. 'Scorchers' who exceeded the speed limit were sometimes sought out by enthusiastic policemen, who would even take cover, armed with stop-watches, in attempts to trap them. Many of the cars of the period were open-topped, calling for drivers and passengers to wrap up well, cover themselves with rugs and wear hats, goggles or veils. There were no garages, and car owners would soon employ chauffeur-mechanics, the first in Scotland being John Lawson, who worked for John Murray at Biggar. These men were often required to repair punctures, a slow process, when the newer pneumatic tyres were in use. Petrol had to be purchased in cans from chemists or motor supplies shops in major centres, and for longer journeys, had to be sent ahead in crates. The chauffeur mechanic, even on short uneventful trips, would have to top up vehicles with water, petrol or oil.

If an impression of the early days of 'automobilism' is easy to reconstruct, the underlying financial position during the first few years of Albion's existence is a different proposition. Partnerships, then as now, were not required to file accounts, and these cannot any longer be traced. It is interesting to speculate on the reasons for the firm's constitution as a partnership in its first phase. In 1899, limited liability companies had only just begun to become respectable, and a number of fairly recent financial scandals connected with the bicycle trade were partly responsible for retarding their widespread acceptance. It is therefore possible that Fulton and Murray may have felt that a private limited company would have deterred customers from paying in advance for cars, given that insolvency could result in the loss of deposits. On the other hand, the standard advantages of informality and privacy may have been dominant in their thinking.

Murray and Fulton contributed £2,000 each to establish the firm, Murray's share coming from a bank loan secured on Heavyside Farm, his father's property. It would be wrong to omit to mention Murray senior's complete commitment to the new business. He had been the second person in Scotland to purchase a motor car, a Panhard, and is likely to have been the driving force behind his son's involvement. The year 1900, which saw the sale of one vehicle only, must have shown a loss, and, as agreed between the founders in their partnership deed, the capital would have been restored to its original level. Interest at five per cent per annum was added to the partners' capital accounts half yearly, and each took one pound ten shillings as a weekly salary. 21 chassis were sold in 1901, which is likely to have eased the financial position. In comparison with this relatively modest introductory capital base, Hozier Engineering, later to become Argyll, started out with £8,785 in ordinary share capital in 1900, with investment rising to £20,000 in 1901 and £45,000 in 1902, attracted by handsome profits and backed by a network of businessmen. Mo-Car was in an even stronger financial position, having been started with £40,000 in share capital, mostly provided by the Coats family of Paisley and Sir William Arrol. Unfortunately for the latter firm, its Camlachie premises were gutted by fire in 1901, but after a year's gap, it would restart operations in a disused Paisley thread mill in 1903.

On 7 February 1902, John Francis Henderson BSc, who had worked with Murray at Mavor and Coulson, joined the Albion Company. By this point production and orders were going well, so well that managerial energies were stretched to the limit. Fulton was tied up with production matters, Murray with the design and product engineering side, and the bodywork, spares and repairs

1.9 John Lawson, chauffeur to John L. Murray, Biggar at the tiller of a dog cart, with N. O. Fulton alongside, 1900.

aspects of the business now required senior attention. It was also increasingly obvious that the Finnieston premises were becoming inadequate. These had a floor space of only 3,600 square feet and remarkably, were on the second floor, making it necessary to move bodies and chassis up and down on a lift. Capital was required to take the company forward out of these cramped circumstances, and Henderson was able to produce a substantial amount. On 5 July 1902, the partnership was converted to a private limited company, in which Murray, Fulton and Henderson each held shares valued £3,750, the original two partners transferring their capital stake into shares and Henderson providing cash. At this stage each of the three became a 'joint managing director', but with Murray as chairman.

The scene was now set for a move to larger premises. By this stage, sales were increasing and staff numbers stood at 43 in the works and five in the office, an increase of 14 and two respectively over 1901. The trustworthy performance of the Albion vehicle was now beginning to attract wider notice. In the 535-mile

1.10 John Lawson and Mary G. Murray (front car) with Mrs T. B. Murray and Mrs N. O. Fulton (rear car) at Heavyside, with John L. Murray standing.

reliability trials associated with the Glasgow International Exhibition of 1901, Walter Creber of Barrhead, who had taken delivery of his dog cart on 8 March that year, won the silver medal, under the discerning eye of a distinguished panel of judges that included Sir John H. A. MacDonald, the Lord Chief Justice Clerk, who was one of the most distinguished Scottish proponents of the motor car. The company now stood ready to step up to a new level of success.

CHAPTER TWO
On the Road: From Cars to Commercials

The probability that the new business might expand rapidly had caused Fulton and Murray to enter into a rental agreement for the Finnieston premises that gave them some flexibility. An annual rental of £65 was payable for five years, but the partners negotiated an optional termination after the third year. This they had no choice but exercise. In anticipation, they had purchased two and three-quarters acres of ground at Scotstoun, several miles further west of Finnieston and close to the north bank of the Clyde.

The parcel of ground purchased was a greenfield site on the edge of the built up area, but had a railway siding. In addition, it was near the engineering businesses of Clydeside and the Glasgow conurbation, from which Albion's labour force was drawn. The factory was ready in July 1903, and consisted of a two-storey south-facing red brick office building of a plain design, the architects, Brand and Lithgow confining their creative contribution to a few classical touches that would not add greatly to the cost. Further along South Street, a single-storey workshop in the same style was erected. This had north-facing roof lights, with pilaster strips suggested on the exterior, and it extended northwards for nine bays. Inside, the unobstructed space allowed for flexible production. The modular design of this workshop allowed for easy extension of the factory as the business grew. This would in due course take place in a westwards direction, filling up the gap between the workshops and the offices in 1905. The first portion of the factory cost £5,000 and a further £2,500 was spent on machine tools, many brought in from the United States.

As one commentator has noted, the erection of a new factory on the outskirts of Glasgow was typical of best practice among the early motor manufacturers. In Scotland, only five new motor works would be built in the industry's early phase. This was a logical second step for rapidly expanding companies. It allowed for unencumbered, rational planning of production, which had not always been possible in the adapted premises they had first occupied, and this advantage was deemed to offset the capital cost of new premises. Hozier Engineering would in due course re-form itself as Argyll Motors in 1906, leaving behind its first home, a converted textile works in Bridgeton, for a new purpose-built factory at Alexandria, also on the outskirts of Glasgow, doing so for the same reasons as Albion. Mo-Car, now called Arrol-Johnston, where a fire caused a transfer to a converted textile works in Paisley in 1902, would in 1913 transfer to a new factory at Heathhall, Dumfries. On a smaller scale, Halley of Glasgow, who had started

2.1 Albion's new premises at Scotstoun, 1903.

up at Finnieston slightly later than Albion, would in 1907 build a new factory at Yoker, not far from Scotstoun, and over in Edinburgh, a new motor works had been built for Madelvic as early as 1899, at Granton. Albion was therefore one of the first to take this important step.

It is no longer possible to state with precision what happened inside the new factory in its early years, but it is recorded that the disruption associated with the shift of premises led to a reduction in the number of chassis produced from 35 in 1902 to 32 in 1903. The factory included a machine shop, smiths shop, coppersmiths shop, assembling and erecting shop, a cleaning shop and finishing stores.

The *Glasgow Evening Despatch* of 30 May 1903 carried a report on the new plant that contained a description of its intended working:

The entire works are on one floor and are airy and well ventilated and cool — remarkably so for an engineering shop. The arrangement, to put it simply, is on the plan of the American sausage machine — inverted. Instead of the live organic animal walking in at the one end and coming out raw material, the 'raw material' goes in at one end of the works and at the other comes out a panting new 'Albion' motor car. In the interval the iron and steel, copper and aluminium, have been planed and pared, turned, drilled and polished, beaten out by pounding hammers, and otherwise put into shape for their business in life.

In the machine shops are the latest productions in the shape of 'tools'. One is steadily paring down a steel plate to the required thickness, doing the work automatically, stroke by stroke. Another is gaily sawing the end of a steel bar nine inches in diameter, which is to be made into a bearings part. In an adjoining room the pneumatic hammer is rapidly beating out a red hot metal bar to an eighth of its original thickness. The floor of the machine shop is laid with concrete, overlaid with asphalt, to make it thoroughly damp-proof, over

17

this again being the white pine flooring and a maple top laid at right angles to the pine planking. The tools can be readily moved from point to point on such a floor without special preparation for their reception. A portion of the machine room is divided off and used as a room for storing the tools and for the making of new tools and 'jigs'.

Adjoining the machine room are the smith's shop and the coppersmith's shop, the latter equipped with brazing furnace and blow pipe for doing all the copper and pipe work required on the cars. Then there are the assembling and erecting shops, fitted out with overhead travelling cranes, and the testing and repair shops. The latter are separated from the rest of the shop by a fire-proof wall. They are also provided with a travelling crane. Beyond is the cleaning shop and a store for finishing cars.

Among the suppliers to the new factory were S. Stevenson and Company of Polmadie, who produced solid tyres, Delta Metals of London, who provided brass frames and Kennedy's of Dolphin Works, Pollokshaws, supplying jigs and fixtures. The firms supplying aluminium parts were A. Brown and Company and the Aluminium Castings Company. Road springs were supplied by Jonas Woodhead of Leeds, coil springs by L. Sterne and company, and engine castings by Boyd. Bodies, supplied by Penmans of Dumfries and Alex Cree, were fitted at the factory. This sample of key suppliers makes it plain that at this stage in the new industry's history there was little need to go outside Scotland for components, and in any case, the English Midlands was not yet established as a specialist source of motor industry supplies. It is clear, too, that the industries of the west of Scotland such as steam engine, marine engine and gas engine manufacture and repair, ensured that there were ample mechanical engineering skills of the right kinds available to Albion and to the young motor industry generally. These facts give the lie to superficial theories that the early Scottish motor industry could not have succeeded on a grand scale as a result of its allegedly 'alien' character compared with the heavy industries of Clydeside, and its distance from component suppliers.

Once firmly ensconced in Scotstoun, technological development was resumed. A two-cylinder vertical engine of 12 h.p. had been designed in 1903, together with a separate 3-speed and reverse gearbox which incorporated a bevel-driven differential linked by two separate chains to the back wheels of vehicles. 12 of the 33 vehicles produced in 1903 were of this type. As demand for more power, to give higher road speeds, increased, the bore of the engine was enlarged. In September 1904 it was increased from $4\frac{1}{2}$ in. to 5 in., and the machine rated at 16 h.p. This design gave reliable performance for many years, forming the foundation on which Albion's reputation for reliability was based. As ever Murray was deeply involved, again following rather than creating trends, and Fulton was for his part ensuring that all production parts were the subject of engineering drawings that gave dimensions and tolerances. Limit gauges were in use throughout the plant, as were fixtures when the opportunity allowed, thus

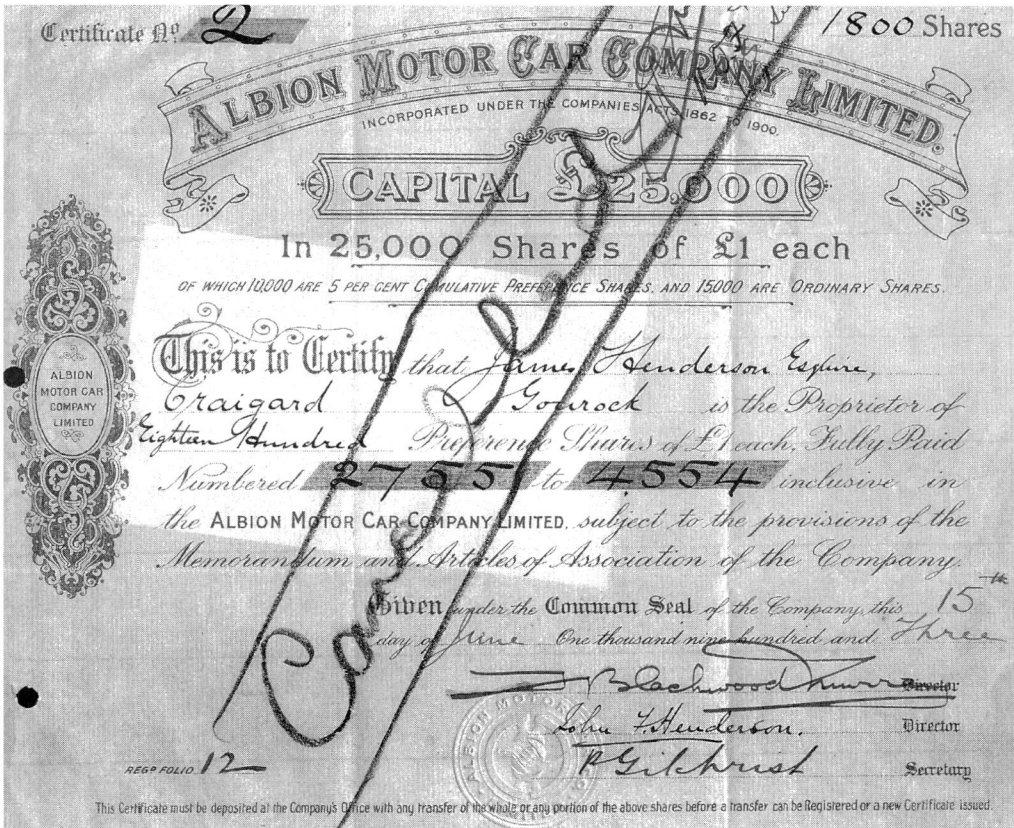

2.2 Cancelled Albion share certificate, 1903.

ensuring that the parts were interchangeable. From this new base, the company continued to prosper.

The British automobile market in the early years of the century was characterised by a preference for foreign cars, especially French models. In March 1903, William Weir, addressing the Royal Scottish Automobile Club, noted that while about 2,500 cars a year were being built in Britain, about 3,500 were being imported. It seems obvious that track record, and therefore reliability, was one major factor which lay behind this statistic. In the face of these realities, Albion's sales and marketing strategy stressed two complementary facts: that every step had been taken to ensure that the vehicles were simply and reliably designed, and that they were British. The very name, Albion, had been adopted to accentuate this latter point.

As well as emphasizing these facts in the publicity brochures that were produced annually at this time, practical steps had to be taken for selling and distributing the vehicles. This involved the appointment of concessionaires in key towns and geographical areas both at home and abroad. Concessionaires were paid from seven point five to ten per cent on sales. Penman of Dumfries, bodybuilders to Albion, entered into an agency agreement at an early stage.

The Albion board minutes record that at various times during the period 1903–1910 the following agents/concessionaires were selling Albion vehicles:

John Love : Kirkcaldy
T. M. Mackay : Ayrshire
Alex Naught : Dingwall and surrounding area
Lage Bros : Rio de Janeiro
G. Bruce : Aberdeen
W. Crockart : Blairgowrie
Sutherland Motor Traffic : Sutherland
Macrae and Dick : Inverness-shire
D. Munro and Sons : Elgin and Nairn
A. Donaldson & Company : Edinburgh
Croall and Croall : Kelso
J. M. Scott : Melrose
George Thomson : Stirling
A. D. Kennedy : Haddington
J. G. Yates : Brisbane
Ayr County Motor Co. : Ayr
Rossleigh Cycle : Dundee
John Colclough : Dublin
D. Miller & Son : Crieff
R. Garden : Tongue
John Rokison : Belfast

The position was dynamic, however, and it is possible that not all concessionaires are recorded. As agreements from time to time proved unfruitful, agents were removed or replaced by Albion.

This left the English market to be considered, and London in particular. In 1905 London had 23.6 per cent of motor vehicle registrations in Great Britain. Discussions were held with Claude Browne, the chairman of Lacre (an acronym for the Long Acre Car Company, which was named after a London street in the neighbourhood of Covent Garden) regarding an agreement. In initial discussions, a radius of 50 miles from London was suggested as a territorial boundary and talks were restricted to the supply of 'commercial chassis'. It was finally decided that Lacre should become concessionaires for the whole of England and Wales, and that the agreement should also apply to Albion cars. Cars sold by Lacre were to be sold as Albions, but commercial chassis bodied and sold by Lacre were to be sold under the Lacre marque, with the proviso that the Albion name had to be displayed on the engine. The normal Albion terms of payment were one-third cash with order, two-thirds when vehicles were complete, but the Lacre agreement was for cash on delivery. Offsetting this concession was the fact that Lacre would take five chassis per month minimum, an indicator of the buoyant state of the London market. In return for this steady business, discounts ranged from 15 per cent upwards, reaching 25 per cent if sales exceeded 50 units per year. An agreement

2.3 10 h.p. tonneau at Scotstoun, 1903.

was signed at the end of 1904 for a five year period. A further significant step taken by Albion was the opening of a Glasgow showroom in 1907.

In addition to these practical arrangements, Albion came to participate in motor shows based in Scotland and in London, at Olympia or Crystal Palace, where the public could be given expert information on the various models on offer. Perhaps surprisingly, Edinburgh had an annual 'Motor and Cycle Exhibition' which had been running since 1898. These shows were widely reported in the motoring press, which included the journals *Autocar*, *Motor Traction* and the Scottish-based *Motor World*. *Autocar* and *Motor World* were extensively read both inside and outside the motor trade. These periodicals would come to serve Albion well in two respects. Advertisements were placed there, and when new models were produced, test drives were taken by journalists and formed the basis of often lengthy articles on the vehicles. The combined effect of these measures was to take Albion chassis sales up from 33 in 1903 to 65 in 1904, rising to 221 in 1906 and 248 in 1907. In 1910, 282 chassis were sold, a total that would go on rising for years to come.

The above production and sales achievements were enacted against a background of change in the detailed design of vehicles. As was noted in the last chapter, the steering wheel superseded the tiller at an early stage, and Albion followed the trend. The gradual move away from solid rubber or steel tyres called for compensatory changes in chassis and brakes, and there was a tendency away from coach-based body designs to styles specially developed for the motor cars and commercial vehicles that were becoming more common. In addition, the

21

2.4 A dog cart in use at Biggar. W. Ovens and boys collect for a bonfire c 1903.

alterations to engine, gear box and drive mechanisms already outlined were made, the consistent elements being Murray's patent engine peripherals. One major departure from this steady, evolutionary pattern requires detailed treatment. In 1906 a 24/30 h.p. car, the A6, was developed, possibly on the advice of Lacre, who had passed on the information that four cylinder engines were preferred in the London market.

The A6 was the first and only Albion vehicle designed solely as a motor car. All other chassis produced up to this point could be fitted with either a motor car or any other kind of body. A report on a trial run in the *Autocar* of 4 August 1906 concluded that 'the 24 h.p. Albion will bear comparison with anything of its class and price'. In 1907 it won a silver medal in the 'Vapour Emission Trials' of the Royal Automobile Club, and in 1908 it scored a further success when it won the Scottish cup for best petrol consumption in the Scottish Reliability Trials. In spite of these positive developments, only 57 units were manufactured, and it was in due course discontinued.

Even before the move to Scotstoun, the composition of the British automotive market was beginning to change, thanks to the development of motorised transport. This was seen early as having the potential to replace the horse. An Albion chassis had been fitted with a van body as early as 1902, but sales of this type of vehicle were not substantial until 1906–7. These were watershed years in the history of the commercial vehicle in Britain. *Motor Traction*, an influential

2.5 John Francis Henderson and wife in an A6 Albion at Cairn-o-mount trials, c 1906.

journal set up to specialise in 'motor vehicles for business purposes', initiated, in 1905, an annual survey of vehicular traffic crossing Putney Bridge on a chosen day. The first statistics revealed that 84 per cent of all traffic was horse-drawn, but by 1906 this had been reduced to 55 per cent, indicating a trend that was to continue steadily. The survey had included private as well as commercial traffic, and took in steam-powered vehicles and motor cycles. Its conclusion was plain: the rising number of motorised omnibuses, taxicabs, vans and lorries was playing an increasingly significant part in replacing the horse.

Motor Traction also gave details of a speech made by J. W. Roebuck, a prominent engineer, on the advantages of 'industrial motor vehicles' over horses, trains and trams. He made a list of 24 advantages that the motor vehicle had over the horse. It included the sanitary and cost factors as well as the highly contestable attribute of not dying suddenly! Arguments such as these would serve motor salesmen well for a long time.

Roebuck's principal objections to trains and trams concerned their high capital cost and operating inflexibility. Trams were beginning to display particular problems: they caused traffic jams, especially when they were broken down. In 1906, a commentator in Glasgow exemplified the point by stating that the number of trams had 'about reached the practicable limit'. He suggested that a supplementary service of omnibuses was needed. There were objections to the seemingly irresistible rise of the motor vehicle. Many resented its noise and smell, and local authorities, who had invested vast sums of money in tramway systems,

2.6 John Buchan and wife in rear of dog cart at the Peebles electoral campaign, 1903.

feared that they were being jeopardised. In 1907, 176 out of the 248 vehicles sold by Albion were for commercial use. Although further details of this kind are not available for subsequent years, it seems likely that this proportion was typical of the rest of the new century's first decade.

Much of this had depended on Lacre. Between 1905 and 1909 about 40 per cent of all Albion sales came through its London-based distributor. In September 1906 relations with Lacre started to go sour. They complained that A. C. Penman had sold some vehicles to a customer in the north east of England, Shipstone, a possibility that had been overlooked when the agreement between Lacre and Albion was framed. Since as early as 1902, Albion had been selling small but increasing numbers of chassis to overseas customers. Lacre's entitlement, or otherwise, to make export sales had not been countenanced during the original negotiations. Lacre also took the view that Albion should not be claiming in its advertising that it was the maker of Lacre vehicles.

In terms of the agreement in place between the two companies, the dispute was referred to a neutral arbitrator. A Cardiff solicitor was chosen early in 1908. After a prolonged inquiry, he assessed damages of £5,000 as a result of Albion advertising the fact that it manufactured Lacre vehicles. A further £6,000 was awarded when it was discovered after the mutual inspection of records that Albion did not manufacture Lacre orders in strict rotation. For the loss of overseas sales

24

2.7 16 h.p. Albion van c 1905.

Lacre was awarded a further £2,750. It was no consolation that Lacre had originally claimed an absurd £62,000. Shocked and dissatisfied, Albion referred these matters to its lawyers. In April 1909 the £5,000 award was set aside by the Court of Appeal and, later that year, the award of £6,000 was overturned. An appeal by Lacre to the House of Lords a year later resulted in the final dismissal of all claims except the outstanding £2,750, whereupon the matter concluded, as the agreement with Lacre had done in 1909. It is difficult to overestimate the pressure that the dispute put on the Albion directors, who must have been concerned for years as to how it might end up. A letter from Murray to his father-in-law Mr. Rusack, the St. Andrews hotelier speaks of the 'awful incubus... never knowing when it might crush or maim us'. If the sums awarded by the Cardiff arbitrator had been made to stick, it could have ruined the business, or forced Albion to cede a controlling interest to Lacre, which is what Browne had wanted all along. He had unsuccessfully tried to negotiate a place on the Albion board in 1906.

The Lacre incident cannot in retrospect be seen as all bad. Between 1905 and 1909 Lacre disposed of some 500 Albion chassis, bodied and sold as commercials by Lacre, throughout England and Wales. Albion itself, over the same period, only managed to sell two-thirds this amount of commercial vehicles in its natural

2.8 Albion/Lacre van in Harrods livery c 1908.

markets. It was universally known that Albion manufactured Lacre chassis, and furthermore, Lacre's customers had been the blue-chip enterprises of London, ranging from Harrods to Bryant and May and Birds. After Lacre had departed from the relationship, these businesses continued to trade with Albion for many years. It seems, too, that Albion's directors, who were particularly hard pressed before the Lacre concessionaire agreement, could not have done so well themselves in distant markets they did not understand. On balance, the relationship was of benefit to Albion, since it spread the company's reputation faster and wider than its own human resources are likely to have allowed. Albion surely deserved this outcome, since by general agreement it had not been guilty of acting out of 'mala fides'. It is certain, however, that Lacre was responsible for the near failure of the A6 car. There is no doubt that it lost interest in it when the disputes commenced, and it was therefore denied access to the London and wider English markets. A board minute in 1912 records the final financial costs of the Lacre episode, which were as follows:

Total legal costs	£11,778-19-4
Plus:Damages awarded to Lacre	£2,750-00-0
	£14,528-19-4
Less:Legal costs recovered from Lacre by order of the court	£3,364-18-4
TOTAL	£11,164-1-0

2.9 John Lawson at Heavyside with Albion 16 h.p. tonneau, 1906.

A footnote to the Lacre episode concerns R. W. Wilson, who had joined Albion as a salesman by 1903. He had left Albion to join Lacre in 1907. In 1909 Wilson, who held a small stake in Albion, wrote to the board asking for copies of the annual accounts, doubtless to pass to Browne. While the Albion board recognised that they were probably under an obligation to provide these, they replied through their solicitors that they could not see their way to meeting the request. This little episode serves to show the extent to which Lacre's success in its markets had depended not just on Albion products, but also on Albion expertise.

During and beyond the Edwardian period, the company had become larger and more complex. Fulton, Murray and Henderson found themselves under immense pressure, which clearly had implications for the company's organisation. Few details of this survive, but it is plain that by 1903, a structure was beginning to emerge. R. W. Wilson, as was noted, was in charge of sales, P. Gilchrist was Company Secretary and doubled for a time as Cashier, J. Gibson was head draughtsman, and A. Rough and David Keachie were foremen, in charge of production. Keachie had been Albion's first foreman, from the Finnieston period. By 1905, the Repair Shop at Scotstoun had its own foreman, J. Millar, but by 1910 would have its own manager, Robin Thomson. A board minute of 1 November 1906 records the following:

The directors having found that the magnitude of the purely commercial work in connection with the control of the business was such as to prevent proper

2.10 Sir T. D. and Lady Gibson-Carmichael at Melbourne in Governor's Albion Landaulette, 1910.

attention being given to the technical side, they decided that it would be advisable in the interest of the Company to bring in another prospective Managing Director whose duties should be that of Commercial Manager.

The appointment of Hugh Ernest Fulton, brother of Norman Osborne Fulton, whose general business background was deemed suitable, was duly made, and immediately he took charge of sales. This included the already vexed relationship with Lacre, which took up an increasing proportion of his time. In 1907, G. M. Young of the drawing office took over from Wilson in sales. In 1909 T. L. Webster was appointed over Gibson as head draughtsman, but it was made clear that his responsibilities did not extend to the design section, which Gibson may have retained. The volume of work had made it impossible for Murray to be involved to the same degree as before in this area. The pattern was evolutionary and the organisation was functionally based, as might be expected of a company started by two engineers. In 1907, Keachie had become production manager, and other departmental heads would emerge as individual departments grew too cumbersome for any single person to manage properly.

It should be emphasised, too, that this ad hoc approach to structure related to the company's close-knit nature. Murray and Fulton had started out as great friends, and the relationship had been deepened through business and inter-marriage. Fulton's wife was one of Murray's five sisters. Fulton's brother had joined the board, and Henderson, before becoming a director, had been a friend and associate of Murray. Personal and business bonds were extended outwards by the directors to senior employees, whose competence, long service and loyalty were often rewarded by bonuses and promotions as the company expanded. This helped create a culture akin to an extended family in every direction.

Some reflection on the personalities of the directors is also revealing in this regard. Judging by their portraits and by surviving reports, the Fulton brothers appear to have been extremely gracious, modest, open-minded and considerate men, excelling in human relations as well as in the technical aspects of their positions. Murray comes across from his surviving correspondence as more complex: open and tender in his dealings with Fulton and with his father and loved ones, but peremptory and autocratic in his relations with outsiders, especially if of a lower social caste than he perceived himself to occupy. His surviving correspondence does relate to his youthful years, however, and the impression is given from later portraits and board minutes of a mellowed personality. No doubt a country boy from Biggar, thrust quite quickly into success and prominence, had adjustments to make.

Murray's more tender side emerges from an interview given to the *Scottish Review* in 1908, in which the happy boyhood relationship with his father is brought out:

> My father had a natural bent [for engineering]…and in those days kept a small well-equipped workshop for experimental purposes and for amusement. In my boyhood he was wont to make me mechanical toys in the place, and being an only son, I was a good deal with him…

The same article opined that Murray was 'probably the most important member of the directorate of the Albion Company'. Without underestimating his technological primacy, it seems likely that Fulton was at least as important in his relations with the men who made the vehicles and who were therefore under his direct charge.

While it is inconceivable that a Clydeside workforce from this period would exist without trades unions being involved, Albion's experience, judged by its surviving records, appears to have been free of difficulties during its first decade. As the critical mass of the workforce grew (from seven in 1900 to 536 in 1910), unions must have gradually made their presence felt, and the directors appear about 1907 to have considered joining the local branch of the engineering employers federation, judging by a newspaper report on this organisation kept in a scrapbook belonging to Murray. A number of factors conspired to help in the preservation of industrial peace during this era. Albion attracted its workforce by offering above-market wages at this stage in its evolution, and the years 1906–1907 were characterised by an industry-wide recession which did not affect Albion and perhaps encouraged employees to 'count their blessings'. Apart from these factors, a number of small gestures doubtless helped keep goodwill. Donations to the employees' Infirmary Fund and to local funds 'for the relief of distress' combined with 'Territorial Camp' allowances for apprentices and occasional extended sick pay for chronically ill employees to portray an image of humanity. As far as can be judged today, the company was a good employer.

Something of the works' atmosphere can be gauged from the recollections of John Keir Ovenstone, who was an apprentice in 1907. He recalled that the door between the Fitting Shop, where chassis were completed, and the Repair Shop, where final adjustments and tests were made, was a bone of contention between employees to the extent that humorous messages were chalked on it to encourage users to keep it shut. These ranged from 'Shut the b___ door' to 'Don't open this door, crawl under'. On another occasion, one apprentice displayed his sense of humour by carrying out an imaginary and melodramatic tightrope walk on a line chalked on the floor, drawing the laughter of operatives in the vicinity. A private talking to from Millar, the foreman, wiped the smile from his face for several weeks. Hard and good work was expected, and one incompetent employee who had incorrectly installed clutches in a batch of vehicles was sacked on the spot. Companionship developed between apprentices and journeymen, and one of these, an ex-navy engineer artificer, had shared a vessel with Prince George, later to become King George V, describing to the trainee how the future monarch was given no preferential treatment whatsoever. The atmosphere appears to have been one where good humour and discipline managed to co-exist.

The Albion finances over the first seven years of Scotstoun trading, predictably in the light of the constant sales growth, were healthy. Profits in 1903 of £751 rose to £2,762 in 1904, £7,070 in 1905 and £11,637 in 1906. Less good results of £7,856 in 1907, a recession year in the industry, and £4,792 in 1908 may reflect either pressures on margins brought about through competition, provisions made for the Lacre lawsuits, or both. 1909 produced a profit of £9,244, and 1910 £21,846. These performances were based on moderate pricing. A surviving list from 1906 gives Albion and Lacre prices at £365–£395, in the middle of the range. Profit as a percentage of issued capital was always satisfactory, ranging from five per cent in 1903 to 37 and 35 per cent in 1905 and 1906. Issued capital was increased very substantially as premises were expanded in 1908, producing returns of seven, 14 and 32 per cent in 1909 and 1910.

The company had no difficulty whatsoever raising extra capital during this phase of its existence. The board minute books record the regular receipt of unsolicited letters from wealthy businessmen seeking equity stakes in the company. These inquirers were frequently informed that a small proportion of what they requested, say 20 per cent, could be given as equity with the balance in preference share capital, an offer that was invariably accepted. This was part of a careful policy. The intention was to keep the vast majority of the ordinary shares in the hands of the four directors, and for external capital holdings to be predominantly in the form of the non-voting, six per cent coupon Preference Shares. Control was extremely important to Murray, the two Fultons and

Henderson, but so too was a desire to keep equity and fixed coupon capital in balance, so as not to overburden the finances with fixed dividend payments in the event of a trade depression. In train with this strategy, calls on share capital were only made when cash was required for expansion, with plenty left in reserve.

As a result, issued ordinary share capital of £12,308 in 1903 rose steadily to £34,967 in 1910, while issued Preference Capital began at £2,604 in 1903 and rose to £32,854 in 1910. About thirty shareholders were involved at this time, and these included (until his death in 1908) John Lamb Murray, Jackson Millar, various friends and relations and a number of local businessmen whose principal holdings were in preference capital.

So far as cash flow is concerned, the figures are no longer available, but as was emphasized earlier, terms of payment were cash on delivery or better, which was likely to ensure little, or no, bank borrowings. The Clydesdale Bank had become Albion's bankers in 1902, taking over from the Union Bank of Scotland, and granted an overdraft facility of £3,000 which does not appear to have been used or extended up to 1910. It was backed by the personal guarantees of Murray, Fulton and Henderson.

So far as the organisation of accounting activity at Albion is concerned, Gilchrist the Company Secretary was by 1910 finding it increasingly difficult to act as both Secretary and Cashier (which may have been the senior accounting post at Albion at this time), and shortly afterwards gave up the secretarial work to Alex Storry, an employee of Carson, Moores, Watson, Albion's auditors. Storry attended to it on a part-time basis from 1912. The presence of a senior cost clerkess implies that junior staff were also employed in this area round 1910, and it is clear from the existence of 'master cost cards' that vehicle costs were being regularly monitored by this staff. The company was as up-to-date in the office as it tried to be on the shop floor.

The decade ended with the threat from Lacre removed, and plans to reorganise the company's English sales operation beginning to be put into action. On 9 December 1909, the former premises of Argyll Motors, at 21–22 Rathbone Place, London, was opened as a showroom and depot, with G. M. Young moving from the sales department at Scotstoun to take over as London manager. Other English depots and offices would soon follow.

Even more important, a board minute of 11 August 1909 discloses the most significant policy change in Albion's history. It records that

> The directors at this date discussed very fully and exhaustively the future policy of the company and after carefully considering the manufacturing capacity of the company and the trend of business, decided to devote their energies more particularly to commercial and utility vehicles.

Both external events and internal preferences had pointed in this direction, and the directors had recognised the inevitable: that for them, the logical way forward was to specialise in commercials alone. No more cars would be designed. Orders for cars would be taken in future, but this side of the business would be allowed to die.

CHAPTER THREE

Expansion 1910–1914: Van Makers to the Empire

The post-Lacre years saw the extension of Albion's English representation, for which £4,000 was provided in the 1910 accounts. The first venture outside London involved the opening in 1911 of a sales office in Sheffield, in Campo Lane. It would take some time for momentum to build up in this part of the world, but by 1913 the company was able to cite a number of Sheffield owners of vehicles in its publicity literature. These were: Duncan Gilmour and Co. Ltd., who were in the liquor trade, and W. D. Yeomans, W. Gunstone and Sons Ltd. and S. H. Jones and Co. Ltd., who were in the grocery business. Added to these were T. B. & W. Cockayne Ltd. and J. Walsh Ltd., described as being in the 'Drapers and Warehousemen' category. In 1912, a Manchester Office and Depot was opened in Deansgate, and in 1913 a dozen miscellaneous concerns, ranging from retail to industrial, were cited as having purchased indeterminate quantities of Albion vehicles. These and neighbouring owners now had access to repair and servicing facilities provided by the manufacturer. In 1914, a Liverpool office was opened on the eve of the war, but is unlikely to have expanded rapidly in view of the government's wartime priorities. It would appear that these offices were started up to generate new business in areas where Albion was not strong, rather than simply to devolve existing business originally taken by Scotstoun.

Orders were generally booming over the 1910–1914 period, with commercials, inevitably, dominant. Sales rose from 282 chassis in 1910 to 354 in 1911, an increase of about 25 per cent, and from 1911 to 1912 they increased to 493 units, that is, by 39 per cent. From 1913 to 1914 the increase was a less spectacular 12 per cent, which may have been associated with the factory extension programme of these years. This had involved the enlargement of the machine shop, in particular, but in 1913 the first stage of a four-storey ferro-concrete block to the rear of the plant was in the course of erection. Quite obviously, sales success was driving up the requirement for production capacity and indeed, the need for a night shift was recognised, and this had commenced in 1910, more or less permanently. The workforce increased from 536 in 1910 to 940 in 1913. These figures include staff of 118 and 196 respectively. On 8 November 1911 the company joined the Engineering Employers' Federation, the local branch being known as the North West Engineering Trades Employers' Association.

Several external factors can be pinpointed as reasons which at least partly explain the rapid expansion in sales, activity, manpower and facilities at Scotstoun. The first is the move away from the horse. *Motor Traction's* annual surveys revealed that horse drawn transport crossing Putney Bridge on the appointed day had declined from 55 per cent of the total in 1906 to 31 per cent in 1909. Sir John McDonald, the Lord Chief Justice Clerk, was himself discovered by a senior legal colleague in London counting the relative proportions of mechanised and horse drawn transport in Piccadilly Circus, which, to his satisfaction, were moving in the direction of the motor vehicle. The second factor which hastened this trend was the introduction of a War Department subsidy to owners of commercial vehicles. This was inaugurated in September 1911, and affected new commercial vehicles and vehicles built subsequent to 1 January 1910.

For some time the British and other European governments had operated a system which enabled the military to requisition horses for training purposes or emergency use. In return for an annual retainer, the authorities were spared the necessity of maintaining much larger numbers of horses than they required on a day-to-day basis. Since the motor vehicle was superseding the horse, pressure had been building up both in military circles and amongst the public for the system to be extended to vehicles. Not surprisingly, Sir John McDonald was again a leading advocate of this scheme. As a senior territorial reservist as well as an outspoken advocate of the motor vehicle, he was well placed to make his voice heard, and advocated the case in many public speeches.

The motor vehicle in military use had a humane dimension relative to the horse. Half a million of these luckless creatures had died in the recently concluded Boer War. A proposed registration scheme had been published in 1907, suggesting that an annual retainer ranging from £1 to £2.2s should be paid, dependent on carrying capacity, and that registered vehicles should be available for purchase by the military at written down value plus 25 per cent, after applying agreed depreciation rates.

The 1911 scheme gave an initial subsidy of £8 if the vehicle had an open chain drive, £10 if it had an encased chain (most Albions were in this category) and £12 if it had a live axle. Payments were made for the following two years, and an extra £10 was available if a spare magneto starter was carried. Taken together, these subsidies were equivalent to a 5 to 10 per cent discount on the cost of the vehicle. If the necessity arose, the vehicles would be purchased by the military on the terms originally proposed in the 1907 draft agreement. The depreciation rate was set at 15 per cent per annum. Those wishing to participate had to equip their vehicles with two front and two rear towing hooks of War Office pattern, and the vehicles needed to have side lamps, tail lamp and at least one headlight and generator. The War Office felt that its requirement to inspect registered vehicles was 'not of an irksome nature'.

In December 1913, Albion produced a sales brochure that gave extensive and unprecedented information on its customers, its sales pitch and the scope of its markets. As the only extant document that provides any detailed insights into these matters, it warrants copious quotation. Its front cover bore the following pointed title, picked out in gold:

ALBION

Why the Albion Company are the largest Manufacturers of Motor Vans in the British Empire

Inside, the narrative began as follows:

Fourteen years ago, when other manufacturers were devoting their energies to the making of pleasure cars, the Albion Motor Car Company began to specialise in the construction of commercial vehicles. Situated in the largest engineering centre in the world, with a staff of highly trained workmen, they produced a machine which speedily became known for its simplicity and soundness of construction, its great strength and extraordinary reliability. Today they are the largest manufacturers of motor vans in the British Empire. When, therefore, the Albion is referred to as the premier commercial vehicle in the world, it is no idle claim which is made, but one which can be supported by evidence of the most incontrovertible character. That evidence we propose to consider in these pages.

The above statement undoubtedly oversimplifies the process by which Albion came to specialise in commercial vehicles, and for that matter (slightly dishonestly) dates it at a substantially earlier point than it took place. It is much harder to disprove its main thrust, however, that Albion had become the best commercial vehicle maker in the Empire.

The first part of the 'evidence' adduced concerned the existence of large fleets of Albions. These were given in the undernoted table, which took up page 3 of the brochure:

Some Fleets of Albion Manufacture

The undernoted large fleets of vehicles have been purchased by the respective firms mentioned:

		No. of Vehicles
YOUNG'S PARAFFIN LIGHT & OIL, Co., Ltd	London	73
J. SHOOLBRED & Co	London	48
Wm. WHITELEY, Ltd.	London	28
J. LYONS & Co., Ltd.	London, Etc	26
LONDON & NORTH WESTERN RAILWAY Co.	London, Etc.	12
PRICE'S Co., Ltd.	London	12
MAPLE & Co., Ltd.	London	12
PATENT STEAM CARPET BEATING Co. Ltd.	London	11
Wm. CRAWFORD & SONS, Ltd.	Liverpool, Etc	20
YOUNG'S PARAFFIN LIGHT & OIL Co., Ltd.	Glasgow	21

		No. of Vehicles
SPIERS & POND, Ltd.	London	12
JOHN BARKER & Co., Ltd.	London	9
SOUTH METROPOLITAN GAS Co.	London	10
WHITEHEATHER LAUNDRY Co.	London	8
JONES BROS. (HOLLOWAY) Ltd.	London	8
BATGER & Co.	London	5
SUTTON & Co.	London	9
FRANK BENTALL	Kingston-on-Thames	5
TH. WALLIS & Co., Ltd.	London	4
LIBERTY & Co., Ltd.	London	3
A. W. GAMAGE, Ltd.	London	3
WM. HANCOCK & Co., Ltd.	Cardiff	8
J. BLAKE & Co., Ltd.	Liverpool	29
UNITED YEAST Co., Ltd.	Manchester	4
KENDAL, MILNE & Co.	Manchester	3
MACFARLANE, LANG & Co., Ltd.	Glasgow and London	12
CHAS. RATTRAY & Co., Ltd	Glasgow	14
J. & W. CAMPBELL & Co.	Glasgow	7
COOPER & Co.	Glasgow, Liverpool & London	7
UNITED CO-OP. BAKING SOCIETY, Ltd.	Glasgow and Belfast	6
McVITIE & PRICE	Edinburgh and London	8
EDINBURGH & DISTRICT TRAMWAY Co. Ltd.	Edinburgh	4
SUTHERLAND MOTOR TRAFFIC Co.	Lairg	18

Governments

H.M. WAR OFFICE	STRAITS SETTLEMENT GOVERNMENT
H.M. INDIA OFFICE	DUTCH EAST INDIES GOVERNMENT
AUSTRALIAN GOVERNMENT	AFGHANISTAN GOVERNMENT
NEW ZEALAND GOVERNMENT	NEGRI SEMBILAN GOVERNMENT
TASMANIAN GOVERNMENT	NYASSALAND GOVERNMENT
CEYLON GOVERNMENT	UGANDA GOVERNMENT
SOUTH AFRICAN GOVERNMENT	SIERRA LEONE GOVERNMENT
SOUTHERN NIGERIA GOVERNMENT	GOLD COAST GOVERNMENT

The fleets concerned had been built up by repeat orders, and these, together with the 'high standing' of the firms concerned, constituted strong evidence of the Albion vehicle's 'undoubted reliability and low running costs'. Later in the brochure, a list of 43 firms of Drapers and Warehousemen owning Albions was given, followed by a list of 44 Bakers and Confectioners. This included Peak Frean, Lyons, Carr, McFarlane Lang, McVitie and Price and Crawfords, leading household names. 33 Launderers and Cleaners, including Pullars of Perth were next cited, followed by 44 Grocers, among which were Liptons and the Cooperative

3.1 15 h.p. 15 cwt. van in livery of Ardath Tobacco Company Co. London, 1911.

Wholesale Society. A list of 38 Brewers and Wine Merchants came next, and then a list of 62 miscellaneous owners which included Bryant & May, Birds, Gamages, Bluthner Pianos and the *Glasgow Herald*.

As well as citing 264 domestic users of Albion vehicles, the brochure made full play of the company's overseas success. This had started as far back as 1901. The order book for that period records that chassis were sent to Kuala Lumpur, Sydney and South Africa, and by the end of 1903 perhaps a dozen were in the hands of private individuals in these parts of the world. It is not known precisely how these orders came about, nor are the exact reasons for the export business growing between 1903 and about 1910 available today. It must have expanded dramatically, though, for in 1911 N. O. Fulton deemed it worthwhile to undertake a factfinding tour of Albion's Australian Agents, Kellow of Melbourne, Vivian Lewis of Adelaide, Canada Cycle and Motor of Sydney and Brisbane and Perth Motor House of Perth. On this tour he also visited Grapes and Riley, agents in New Zealand and took in at some stage visits to D. K. Saker of Johannesburg and Benjamin and Lawson of Cape Town. By the autumn of 1913, the Albion Motor Company of Canada was being formed at the instigation of A. H. Jones, incorporation coupled with local ownership being seen by this particular agent as the best form of representation. In the middle of 1913, Murray had visited I. M. Singman, Albion's agent in St. Petersburg.

Export business had indeed exploded, and the most obvious explanatory factor is given in the text of the 1913 catalogue. It referred to the

extraordinary demand of Albions for overseas work. An overseas user cannot afford to run the same risks with regard to his purchase as is possible to a firm situated, say in London, where spare parts are easily available. Repair shops are often many miles away.

Great strength and simplicity of control is essential if the machine is to be entrusted to a comparatively uneducated driver with complete success.

The condition of many overseas roads, too, makes it imperative that Albions should be ordered. Its high ground clearance, its ability to withstand shock and hard use, its low petrol consumption, and its proven great length of life, makes the Albion almost indispensable to the user of commercial vehicles in outlying districts.

Even allowing for the salesman's 'puff', it is difficult to avoid the conclusion that the engineering merits of the Albion machine were substantially responsible for its overseas success.

Pictures of overseas-based Albions were shown in the liveries of Uganda Transport, Leeson, Dickie, Gross & Co. Ltd. of Vancouver and the Ceylon Wharfage Co. Ltd. of Colombo. Something of the flavour of what must have happened to build up business is found in a testimonial letter from the Cycle and Carriage Company of Kuala Lumpur, one of six printed in the brochure:

Dear Sirs — We have now many Albions running in the Straits and Federated Malay States. These cars are for passenger service, and run from 80–120 miles daily, and some of them are on the most hilly places in this country.

The simplicity and reliability of these Albions render them pre-eminently suitable for native drivers, where expert aid is so unobtainable in this place.

The first car was got in 1906, and we are surprised to find that it runs just as well as a new car.

We are well satisfied with the behaviour of the cars in every direction.

A list of 'Some Overseas Users of Albions' pointed out that many governments had purchased Albions. It read as follows:

Governments

Australian Government:
>Printing Department
>Railway Department
>Postal Department
>Tourist Bureau

New Zealand Government:
>Postal Department

Railway Department
Public Works Department
Tasmanian Government
South African Government
Southern Nigeria Government
Negri Sembilan Government
Nyassaland Government
Uganda Government
Sierra Leone Government
Gold Coast Government
Ceylon Government
H.M. The Amir of Afghanistan
Straits Settlements Government
Basutoland Government
British Guiana Government
Russian Government
Dutch East Indies Government

In addition to this impressive proof of Albion's penetration of each of the world's continents at government level, a list containing a selection of customers in Queensland (13), New South Wales (27), Victoria (4), New Zealand (14), Canada (17), Africa (8), India, Afghanistan and Burma (10), Ceylon (5), British West Indies (5), South America (8), and Straits Settlements (6), was given. Perusal of the list gives the impression not only that Albion's agents had been busy, but that the vehicles sold were of every conceivable type, from charabancs to military waggons.

In view of these achievements, the dwindling sales of cars were stopped in 1913. The phenomenal success just described led to excellent financial results in the pre-war period. Profits in 1911 were £21,988, representing a splendid rate of return of 27 per cent on issued capital. Profits in 1912, at £26,608, yielded 28 per cent, and 1913, at £30,132, 23 per cent. The surviving financial detail up to 1914 does not extend to the cash position, but with demand brisk and terms of trading that insisted on cash before delivery, this is likely to have been positive, the one possible exception relating to the fact that the business was constantly extending its workshops, and in 1913 would require cash for the new concrete block and office block that were erected at that time. Any pressures there might have been from this source would undoubtedly be offset by the relative ease with which Albion was able to raise fresh 6 per cent preference capital from private investors. This would have meant that the preference capital could have outstripped the ordinary capital in 1912 and 1913, which would have caused the amount of fixed interest capital to exceed ordinary capital. It was well understood even in these relatively unsophisticated times that this was not a healthy trend, since in periods of low profits, the fixed returns could easily constitute a heavy burden on cash flow, and the Albion directors constantly monitored the balance of fixed and ordinary capital. At the same time they turned reserves (which

3.2 The precursor — Albion's first export, sent to Kuala Lumpur in 1901.

belonged mostly to the directors anyway) into ordinary shares, which were issued to existing shareholders in proportion to their previous holdings. In this way, directors shareholdings and voting powers increased, tightening their grip on the reins of control.

The new buildings already referred to were distinguished examples of their kind and a great source of pride to the Albion directors and employees for many years to come. The concrete block, the eastern portion of which was erected in 1913, became something of a minor landmark in the Scotstoun area, as well as a symbol of Albion's growth and progressive engineering policies. This started its life when Murray and David Keachie undertook a visit to the United States in 1912 to look at the latest manufacturing methods and premises, with a view to extension. A board minute records that they paid special attention to

Ferro-Concrete Buildings and the general arrangement of the various departments. Much valuable information was obtained and after due consideration the Company decided to carry out a further Extension to the works largely on American lines, but with such modifications as appeared desirable and necessary to suit local conditions.

On the 20 December 1912 there were meetings in the U.K. with Albert Kahn, the designer of some of the latest ferro-concrete structures in he U.S., including the Ford Motor company's Highland Park works in Detroit.

Ferro-concrete structures had been evolving in France from about the middle of the nineteenth century onwards. Various systems had been devised for making beams and for fastening together the structural members of concrete buildings, one of the most successful being that of François Hennebique, but the potential of this new type of building was only realised around the turn of the century in the United States, as industrialisation matured there. Research has shown that Kahn was not the originator of this type of building in the U.S., as had been first thought, but an important pioneer with a number of motor car factories to his credit, including the Packard factor of 1905 in Detroit. Ernest L. Ransome (1884–1911), who had designed the United Shoe Machinery Plant in Beverly, Mass., in 1903 was an earlier pioneer, and the architect Frank Lloyd Wright had used ferro-concrete in his modest E.Z. Polish Factory in Chicago in 1905.

The designs for the Albion building were carried out by Kahn through his London-based operation, the Trussed Concrete Steel Company, which almost simultaneously was designing the new Heathhall works near Dumfries for Arrol-Johnston. The Albion block was of four storeys (unlike Heathhall's three), and rose 60 feet. It was 60 feet wide and ran northwards for two hundred feet. In due course (1914) it would be doubled in size. The task of constructing the concrete block was given to S. Stevenson of Polmadie, who quoted £9,000 for this work, the price excluding lifts, glazing, steel, lighting, etc. Stevenson's were the preferred contractors. While their tender had not been the cheapest, their experience at Arrol-Johnston's plant at Heathhall, which they were already constructing, gained them the contract.

At around the same time, a new office block facing towards the Clyde on South Street was being completed. Expansion of the kind Albion had witnessed had also put pressure on the office accommodation, and the solution was to engage Alexander Nisbet Paterson, the distinguished Scottish architect, to design a new office building. Paterson (1862–1947) was a friend of John F. Henderson and lived in Helensburgh. Paterson's brother James was a painter and his wife Maggie was the young sister of the artist J. Whitelaw Hamilton. He was a close friend of several of the Glasgow Boys, particularly Joseph Crawhall, E. A. Walton and James Guthrie, and was an excellent watercolourist and architectural draughtsman. Although he was capable of designing in any style, Paterson's speciality was Scots Renaissance work of the type lately popularised by Sir Robert Rowand Anderson and Sir Robert Lorimer, and his own house, Longcroft, Helensburgh (1901) was in this style. He also designed suave institutional buildings, perhaps the most famous of these being the Liberal Club (1908–9), which stands at the corner of St. George's (now Nelson Mandela) Place and Buchanan Street, Glasgow. In this building Paterson's classical work can be seen at its best. Having been trained under Burnet, Son and Campbell he went on to study at the École des Beaux-Arts in Paris, which specialised in symmetrical classicism almost as an article of faith. This was the style Paterson chose for the Albion offices.

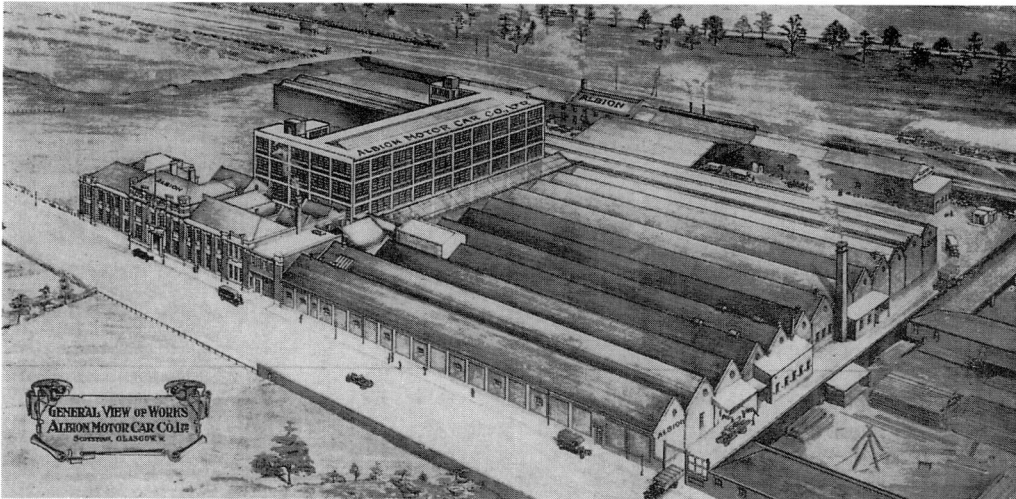

3.3 Albion offices and factory as constructed by 1913.

Built on two storeys, the offices were constructed in red brick with cream sandstone dressings. Doubtless the red brick was intended to harmonise with the workshop frontage to South Street. The central portion of the offices was of three bays, and had an attic storey set behind a balustrade. The middle bay was canted, and framed the twin-columned main entrance door. The four bays on either side of the central section were separated by pilasters, and the building terminated in a pavilion on each side. Externally, the building gave off a sense of classical restraint, but inside, Paterson's Arts and Crafts leanings were expressed with great panache. The Arts and Crafts movement of the late Victorian and Edwardian periods put an emphasis on craft work, and had been inspired by the ideals of William Morris. It was typified by work which took liberties with historic styles, and was close in spirit to Art Nouveau, from which it is at times indistinguishable. Its key feature was a love of the aesthetic properties of materials. At Albion, this was seen in the punched bronze of the light fittings, the etched glass of the magnificent screen just inside the front door, and the glorious marble staircase with its wrought metal baluster.

The planning of the offices was very straightforward. Just inside the front door, on either side, were two tiny offices, designed for secretaries. Adjacent to these, again on either side, were two senior management rooms, probably designed for directors. Beyond these, in each wing, were suites of general offices. The stairs ran to the right of the two-storey entrance hall and gave access to a landing on the first floor. This led to the boardroom, which was above the front door, and on either side of the boardroom entrance was gained to general offices. To the north, access to the works and concrete block was obtained. In later years, Murray's daughter was to describe these offices as his 'pride and joy'. For thousands of Albion workers, they were the company's nerve centre, somewhat daunting, where

3.4 The new Albion office block in 1913.

the hierarchical distinction between management and works staff was all apparent. To customers and visitors they were extremely imposing. In due course they would come to have a symbolic value, somehow encapsulating the company's spirit. Notwithstanding their considerable presence, the offices cost a relatively modest £13,500.

By 1914, Albion had already become something of a byword in the trade, and apart from the Lacre setback, had enjoyed practically uninterrupted success. But what of the rest of the young motor industry, of which it was part? By this point, the industry was already a major force in British engineering, employing about 100,000 people. Daimler alone had 5,000 workers, Wolseley 4,000 and Sunbeam and Humber 2,500–3,000. As Berrick Saul has observed, at this time only the great armament works and shipyards were employers on this scale. Perhaps about twenty U.K. firms at this point made in excess of 500 vehicles per year. The very largest, Daimler, Wolseley, Humber and Sunbeam, had pursued the large scale production of motor cars, and produced 1,000, 3,000, 2,500 and 1,700 chassis respectively in 1913.

The story was very different in Scotland. Out of 50 starters in the industry before 1914, only ten had managed to produce more than 50 vehicles during their existence. Stirling, Stewart and Drummond had managed this total, but by 1914 had gone. Belhaven, which made quite small numbers of cars and lorries in Wishaw, was still afloat and remained so until 1924, as was Sentinel, whose steam vehicle manufacture was to be transferred to Shrewsbury in 1915.

42

3.5 Albions in livery of John Barker & Co. c 1913.

Halley, which had set up in Finnieston in 1901, had subsequently moved to Gallowgate, Glasgow, and then to a purpose-built, mushroom-columned factory at Yoker, where it would survive until 1935, making relatively small numbers of trucks. In 1913, Clyde started up making two vehicles per month in Wishaw, lasting until 1932. The other 40 starters in the industry, such as Madelvic, Grampian, Cassell or Kennedy, either picked the obsolescent steam or electric technologies, or were unable to commit themselves deeply enough to the new industry. This fatal lack of commitment tended to stem from pitifully small capitals, or a failure to move beyond the assembling of bought-in parts and become technologically autonomous.

From fairly early on, it had become apparent to observers of the Scottish industry that only three firms stood any chance of succeeding on a scale sufficient to make a large and lasting impact on the Scottish economy. These firms were Arrol-Johnston, Hozier/Argyll and Albion.

As was observed Mo-Car, which had become Arrol-Johnston in 1905, had just relocated to Heathhall. On its removal to Paisley in 1902, its fortunes had picked up. Johnston himself appears to have retreated into the background by this period, which saw William Beardmore, chairman of the heavy engineering company of that name, come into the business and inject capital. After 1902, the managing director was J. S. Napier, and in 1905, a new car was introduced. By 1906/1907, the company was in a poor financial state, but by 1908, its fortunes picked up. T. C. Pullinger became General Manager, bringing Humber and Darracq experience with him. He introduced new models and sales rose to 300 units per year. In 1911, a large six-cylinder car, rated at 23.9 h.p. appeared, accompanied by a modestly-priced 11.9 h.p., forming a foundation for a successful pre-war period of trading and the decision to expand at Dumfries. Undoubtedly a major factor in its success to date was its financial backing. Mo-Car had started out with £50,000 in capital and its promoters had doubtless to intervene at several points in its history, without which it appears it might not have succeeded. By 1914, it had produced

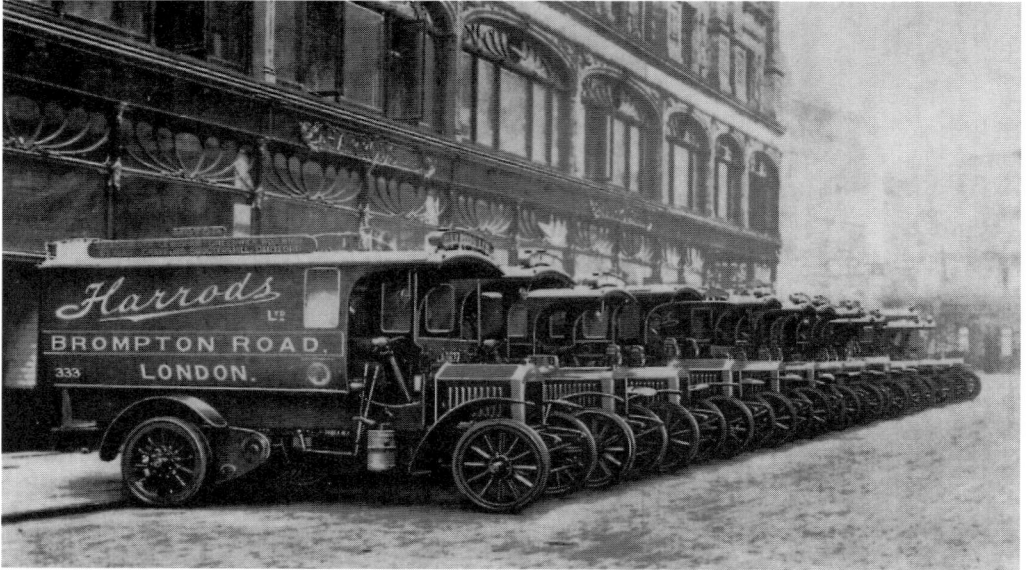

3.6 Part of Harrods' fleet of 70 Albions, 1913.

a cumulative total of about 4,000 vehicles. While this company had had its setbacks, the level of backing it had always received meant that, all things being equal, its chances of success were greater than Albion's.

The *cause célebre* of the early Scottish motor industry was of course the Hozier/ Argyll collapse. This was the firm that, almost alone, could have dynamised the whole Scottish industry, had it fulfilled its potential. Hozier Engineering Ltd had started out as a Bridgeton-based motor car repair works, in the former premises of the Scottish Manufacturing Company, a bicycle business. These were acquired by William Alexander Smith, an entrepreneur, and Alexander Govan, who had worked with the failed bicycle firm. Govan built the first motor car in 1899, based on the 1898 Renault. New cars followed in 1901 and 1902, based on bought-in engines. In 1903, another completely new car was launched, and 30 permutations of body and engine were now available. This time the new car had an in-house manufactured engine. Production rose to 15 units per week, and Hozier became the first firm in Europe to produce at a rate of over 500 vehicles per year, although it is unlikely this annual output total was exceeded at this stage.

By 1905, Govan had organised the Bridgeton factory to work substantially with semi-skilled labour using jigs and fixtures. He too had visited the U.S.A., and Bridgeton was generating handsome profits. The move to a new greenfield site factory beckoned, offering the opportunity to step up production to unprecedented levels of 2,500 vehicles per annum, the highest in Europe, and to completely reorganise the new factory from first principles. The capital raised for the Bridgeton business had risen from £8,785 in 1900 to £62,000 in 1904, three and a half times the Albion total at this stage. It climbed to £244,904 in 1905 and, after its incorporation as a public limited company in 1906, to £474,051. Profits had climbed from £6,109 in 1903 to £26,634 in 1905.

As the new works, rationally planned and with a large and opulent office building, neared completion at Alexandria, Dunbartonshire, profit from the combined old and new operations rose to £46,129 for 1906. Then a number of decisions were made that ushered in tragedy. First, although the shares in Argyll Motors Ltd, as the new company was called, were subscribed for three times over, insufficient capital was issued to cover for the erection of the magnificent new complex and to provide for an ongoing level of working capital. As a result, the company relied on overdrafts from the Bank of Scotland to help with factory building, working capital and trade fluctuations. In addition to this lack of financial foresight, chassis were being produced in advance of orders, contrary to the practice of firms like Albion or Austin in England. Vast stocks were built up in late 1906, in anticipation of ready sales. Instead, a savage trade depression hit the motor car industry, causing heavy price discounting and widespread overstocking. Attempts to reduce pressure by selling share and debenture capital were made in 1907. By 1908, cash flow, weakened by the trade depression and an aversion among investors to Argyll stock, was insufficient to meet fixed costs and debenture interest, and voluntary liquidation ensued.

The company had been reconstructed, with creditors taking huge losses, in 1909, and set out again as Argyll's Ltd. Between this time and 1914, it produced many beautiful new models, but its managing director, Colonel Matthew, who had been seconded to the post by Dunlop, one of the largest creditors, became obsessed with the sleeve valve engine. This was a device that dispensed with valves and tappets and exhausted through its side. In spite of the brilliant engineering work of M. Perrot, who was then in charge of design, reversals in patent lawsuits connected with the new engine cost the company some £50,000, and it finally went into liquidation in 1914. Liabilities exceeded assets by about £80,000 at this point. Who had been at fault? Govan, a brilliant engineer, had died in 1907, but by this point the plant was set up. The vehicles were excellent. The company's financial mismanagement, for which Smith was substantially to blame, is the prime culprit, but the evidence suggests also that the Bank of Scotland was initially too lax with the company.

The contrast with Albion could not have been greater. It had succeeded because it had specialised, and because of its cautious directors, whose business background had been invaluable. It could certainly be argued that the directors' careful policy of retaining control of the company and balancing equity and fixed interest capital held back the company's growth. That may well have been true but it had still prospered, and events lay ahead that would soon give an unplanned lift to its fortunes.

Some of the many Businesses
using Albion manufactures

400 gallon Tank Waggon supplied to Messrs.
Youngs Paraffin Light & Mineral Oil Co., Ltd.

25 cwt. Van, supplied to Messrs. Bryant and May,
Ltd., London.

25 cwt. Van supplied to Messrs. Bluthner & Co.,
Ltd., London, W.

Hearse supplied to Messrs. Attree and Kent,
Brighton.

32 h.p., 5 ton Van supplied to Messrs. Sutton &
Co., London and Manchester.

STAVELEY COAL & IRON CO., Ltd., Coal
   Merchants, ... London
J. L. HENSON, Butcher, ... London
SUTTON & CO., Carriers, ... London
J. H. KENYON, Ltd., Funeral Undertakers,
    ... ... London
GEORGE HARDING & SONS, Ltd., Hardware
   Merchants, ... London
BRYANT & MAY, Ltd., Match Manf., ... London
WANDSWORTH BOROUGH COUNCIL,
   Municipalities, ... London
SOUTH METROPOLITAN GAS CO., Muni-
   cipal Dept., ... London
A. BIRD & SONS, Ltd., ... London
A. SANDERSON & SONS, Ltd., Wall Paper
   Manufacturers, ... London
ALFRED LEVETT & CO., Nurserym'n,London
A.W.GAMAGE,Ltd.,AthleticOutfitters, London
J. DICKINSON & CO., Ltd., Printers, London
BURROUGHS WELLCOME & CO., Phar-
   macies, ... London
ALLEN & HANBURY'S, Ltd., Pharm., London
DUNCAN FLOCKHART & CO., Phar., London
WALTER SAVILLE & CO., Piano Manufac-
   turers, ... London
BLUTHNER & CO., Ltd., Piano Manf., London
C. BECHSTEIN & CO., Piano Manf., London
MIDLAND RAILWAY CO., ... London
STAR NEWSPAPER CO., Ltd., ... London
INDIARUBBER & GUTTAPERCHA WORKS
   Co., Ltd., Silvertown, ... London
LYON & HALL, Piano Manufacturers, Brighton
CHIDZOY & SONS, Fruit & Potato Merchants,
    ... ... Swansea
MILLS, ENGLISH & CO., Ltd., Mechanical
   Engineers, ... Swansea
OXFORD GAS, LIGHT & COKE CO., Muni-
   cipal Dept., ... Oxford
THE SEVERN & CANAL CARRYING CO.,
   Ltd., General Carriers, ... Birmingham
THE ASTON CHAIN & HOOK CO., Ltd.,
   Chain Manufacturers, ... Birmingham
W. H. JAMES, Hide, Skin, Fat and Wool
   Merchants, ... Birmingham
DALE, FORTY & CO., Ltd., Piano Manufac-
   turers, ... Birmingham
SINGLETON & COLE, Ltd , Tobacco Manu-
   facturers, ... Birmingham
SMITH, FAIRE & CO., Boot Factors, Leicester
A. T. VERNON, Butcher, ... Wolverhampton
T. BANTOCK & CO., Carriers, Wolverhampton
J. E. DOWNS, Butcher, ... Bilston
HANS RENOLD, Ltd., Chain Manufacturers,
    ... ... Manchester
SUTTON & CO., ... Manchester
WOOD & SONS (Manchester), Ltd., Manchester
ARTHUR YEWDALL,Dairy Prod., Manchester
RICHARD JOHNSON,CLAPHAM & MORRIS,
   Ltd., Iron Merchants ... Manchester
SIMPSON & GODLEE, Ltd., Calico Printers,
    ... ... Manchester
SADLER & CO., Ltd., Chemical Manufacturers,
    ... ... Middlesbrough
SOUTH SHIELDS GAS CO., Municipal Dept ,
    ... ... South Shields
T. DICKSON & SON, Hide Brokers, Edinburgh
R. W FORSYTH, Ltd., Outfitters, Edinburgh
A. & D. PADON, Printers, ... Edinburgh
EASTON, GIBB & SON, Contractors, Rosyth
JAMES BOWEN & CO., Funeral Undertakers,
    ... ... Musselburgh
BAYNE & DUCKETT, Boot Factors, Glasgow
WYLIE & LOCKHEAD, Ltd., Removal Con-
   tractors, ... Glasgow
ROSS'S DAIRIES, Ltd., Dairymen, Glasgow
MALCOLM CAMPBELL, Ltd , Fruit Mer-
   chants, ... Glasgow
BOWS EMPORIUM, Ltd., Hardware Mer-
   chants, ... Glasgow
R. RAMSAY & CO., Ltd., Hide Brokers, Glasgow
GLASGOW CORPORATION GAS DEPT.,
   Municipal Department, ... Glasgow
BRECHIN BROS., ... Glasgow
GLASGOW HERALD & EVENING TIMES,
    ... ... Glasgow
GOVAN PARISH COUNCIL, ... Glasgow
TAYSIDE MOTOR CO., Carriers, Dundee
NORTHERN CO-OPERATIVE CO., Ltd.,
   Boot Factors, ... Aberdeen
SIMPSON & MORGAN, Boot Factors, Aberdeen
HIGHLAND RAILWAY CO., ... Inverness

25 cwt. Van supplied to Messrs. A. W. Gamage,
Ltd., London.

One of a fleet of 10 Vehicles supplied to Messrs.
The South Metropolitan Gas Co., London.

25 cwt. Van supplied to Messrs. The United
Yeast Co., Ltd., Manchester.

2 ton Lorry supplied to Messrs.
Wm. Pearce, Ltd.

30 cwt. Van supplied to Messrs. Price's Co.,
Ltd., London.

3.7 Page from Albion's 1913 sales brochure.

CHAPTER FOUR

Albions at War: Production Under Pressure

The spark that ignited the Great War was the assassination of Archduke Franz Ferdinand and his wife on 28 June 1914 at Sarajevo in Bosnia. The Archduke was heir-apparent to the Austro-Hungarian Empire, which was ruled by the Emperor Franz Josef. The Empire was under threat from an upsurge of ethnic nationalism. The Austrians accused the Kingdom of Serbia of harbouring the assassins, who were Serbs, and set out on a course of revenge and repression. On 23 July they sent an ultimatum to Serbia demanding that the assassin be found and arrested, and that anti-Austrian propaganda be banned. Deeming the Serb response to be unsatisfactory, Austria declared war on Serbia on 28 July.

A knock-on effect ensued, and country after country came to side either with Austria or Serbia in line with treaties, historic loyalties and tribal and dynastic relationships. Germany sided with Austria, Russia sided with Serbia, and the French, still smarting from their defeat by Prussia in 1870, saw the opportunity to regain the lost provinces of Alsace-Lorraine through victory over Germany. On 1 August, Germany declared war on Russia, and that same day France ordered the mobilisation of her military forces. Sir Edward Grey, the British Foreign Secretary, also had just previously asked both France and Germany if they would observe Belgian neutrality, received an affirmative reply from France and none from Germany, whereupon Belgium mobilised her forces. On 4 August Germany invaded Belgium, and the same day, Britain declared war on Germany, following this on 12 August with a declaration of war on Austro-Hungary.

Germany's military strategy, which had been in secret preparation for several decades, was to swing through Belgium and into France in a giant door-like formation, hinged on Switzerland and moving anti-clockwise: the Schlieffen Plan, named after its (by 1914) deceased author. This was intended to take France by surprise, in the expectation that her armies would be deployed in greatest strength in the area of Alsace-Lorraine and would not be prepared for an attack through Belgium. At the start of the war some one and a half million soldiers formed the offensive 'door'. The initial plan was to suppress France within six weeks and then turn German attentions to the eastern front where, together with Austro-Hungary, she would concentrate on the defeat of Russia, Serbia and their allies.

It was the failure to achieve this initial success that attenuated the war and turned it into the apocalyptic bloodbath that wiped out a generation of Europeans and sealed the fate of the social order of the 19th century. In the east, a gallant raid by the Russians on east Prussia resulted in the rout of the Germans at

Gumbinnen on 20 August. Shortly afterwards, the Germans reversed Russian progress at various points, but the Austro-Hungarian invasion of Serbia on 12 August would within a few months be thrown back. Meanwhile, on the western front, the French had replied to the German onslaught by meeting attack with attack. They thrust northwards into the north of Lorraine at the hinge of the German offensive, but after a week of heavy fighting they had to fall back towards Paris, taking the small British Expeditionary Force with them. Four German divisions were at this point sent to the Russian front, and the German centre and left wheeled towards Verdun. The German First Army, in response to a gap opening up with their colleagues in the Second Army, headed south. In effect, the Schlieffen Plan was abandoned. French counter-attacks followed, and gradually the opposing Franco-British and German armies began to dig themselves in for the war of attrition that would last for four years and that had already claimed so many lives. Henceforward, until the final stages, no attack nor any repulse made by either side enjoyed lasting success.

Millions of men would come to face each other in trenches across a churned up and shell-pocked no-man's land. When assaults were launched, no surprise was possible. A long period of murderous and deafening artillery bombardment, intended to silence the fire of the enemy and break a gap through his defences, signalled the forward march of heavily armed and laden infantry, who were usually mown down in short order by machine gun fire. Dead and dying men would hang for days on the barbed wire between the lines. For much of the war, the full horror of the front-line experience was only known to direct participants.

Preparations for the war reflected the fact that almost a century had elapsed since Britain had been involved in a major European conflict. Technology had moved on since then and the main lessons had been forgotten. There was no experience of the industrial efforts required, nor of the economic effects of a full-scale struggle. The general expectation was that the war would be short, and while Britain's naval strength was undoubtedly superior to Germany's (the result of a 'naval race' that had been in progress for some time), a response to the demands of the land-based struggle and the requirements of domestic life had to be improvised. Initially, the government restricted itself to specific acts of intervention: the Railway Companies for instance, were placed under government control, with management remaining in existing hands, and a Royal Commission on Sugar Supplies was constituted. This commodity had previously come to a large extent from European countries that were now Britain's enemies. A secret reserve of wheat was built up and the government assumed control of India's crops. The War Office, to secure its supplies, began purchasing meat abroad, and ended up having to take responsibility for trying to prevent inflationary price increases in the home market. In due course it also found itself having to purchase the whole jute crop (for sandbags), Russian flax crops and the entire home wool supply, for blankets. In short, government at first relied on private enterprise and the law of supply and demand. This they felt would produce prosperous trade, high and well paid employment and good profits to tax, which would help fund the war.

Unfortunately, this ad hoc way of proceeding did not cope with the sheer magnitude of the war's demands. Each intervention had knock-on effects elsewhere, and led to others. In due course the state found itself directing the major part of the country's industries.

As far as munitions were concerned, the government initially felt that all that was required was to give the Service Departments vastly increased financial resources, leaving them to requisition from their normal suppliers, who would respond to the good prices on offer and deliver what was ordered. This led to acute shortages, and eventually non-standard suppliers to the military had to be used extensively. One example of difficulty in this area was the 'Great Shell Scandal'. A shortage was causing unnecessary deaths at the front. All this led, in March 1915, to the granting of greater powers to government under the Defence of the Realm Act, allowing it to take over any existing factory or workshop it required. This was not sufficient to increase the supply of shells, and in May 1915, Lloyd George was put in charge of the new Ministry of Munitions. Government from this point onwards began to take a firmer control of 'munitions', defined as arms, ammunition, transport, aircraft and optical instruments used by the forces.

The new ministry quickly grew from 20 clerks in the Army Contracts Department in August 1914 to a staff of 65,000 by November 1918. It eventually had direct control of two million workers, a number which rises to three point four million if those indirectly controlled are counted. It ran 250 government factories, quarries and mines and supervised 20,000 'controlled' establishments. Over the period of the war it would disburse £2 billion.

How was Albion affected by these circumstances? Its first intimation of the commencement of hostilities was a telephone call from the police to N. O. Fulton on 4 August to advise him that all vehicles in Albion's hands were being commandeered and were to be prepared for immediate despatch. The War Department then 'impressed' all future production of the A10 32 h.p. model, instructing Albion to increase the output. The company, which had been working to full capacity, decided the time was ripe to expand, and did this by completing the second portion of the concrete block, to the west, which added a 'mirror image' equivalent to the first portion. This was part of an increase in facilities that added 56 per cent extra space to the Finishing Off and Repair Department and 130 per cent to the Machine Shop by 1915. The idea behind the concrete block had been to achieve an efficient flow of work through superimposed workshops, which was obtained.

In response to the appeals of the military for recruits, about a hundred Albion apprentices and journeymen decided to enlist immediately in their country's service. Some joined the Army Service Corps to tend to or drive military vehicles, while others joined the infantry, the navy, the engineers or artillery to make a more direct contribution. To support those serving with 'the colours', a committee was formed at Scotstoun at management's instigation. Details of its operation have gone, but its purpose was to have each member taking responsibility for liaison with groups of families whose men had joined up, providing them with moral support.

4.1 Chassis assembly, 1914.

A large number would not return. Enlistment and the raising of production levels would soon pose increasingly acute manpower crises over the war years, and this problem was to affect much of Clydeside, which was naturally a major centre of munitions production. At Parkhead Forge, for example, Beardmore was busy making huge guns for warships and heavy artillery use and close by, at Yoker, capital ships for the navy. Just along the road, at Dalmuir, it made a range of diesel engines for military use. At all the main shipyards of Clydebank and Glasgow, work was in hand on a whole range of vessels of strategic national importance. Nearby, Halley was making trucks for the army and in Wishaw, the small Belhaven Company was also engaged in this activity. Arrol-Johnston, in its new Dumfries factory, made aero engines. The steelworks and sub-contractors which fed these industries also suffered from the same manpower problems as the industries they supplied.

Within twelve months of the outbreak of war, workers in crafts deemed to be important to the war effort were exempted from enlistment, and the existing stock of unemployed men had been taken up either by the military or industry. The eventual solution that was found was the substitution of skilled labour with unskilled labour, known as 'dilution', which was introduced to munitions firms because of the impossibility of bringing sufficient skilled men out of an apprenticeship system that took several years. In particular, its most controversial aspect, so far as Albion was concerned, was the dilution of male labour with female labour. This step was more easily taken at firms like Albion, whose up-to-date approach to engineering and acquisition of semi-automatic plant meant that many jobs were already semi-skilled. A quarterly management meeting minute of 7 February 1916 records that 'the contemplated scheme of Labour Dilution would, in all probability, ease the situation'.

At the following quarterly meeting on 1 May 1916, it was minuted that the first stage of labour dilution involving women at Albion had gone ahead smoothly:

> Particulars of the arrangements made under the Dilution of Labour scheme were given, Mr. Keachie stating that, in all, 44 women had been employed to replace men in the different Departments the men thus replaced being transferred to a higher grade of work. On the whole, the scheme had been fairly satisfactory, the Machinemen transferred to the Fitting Shop proving themselves quite capable of undertaking the simple fitting jobs, and the women, after a period of careful supervision, proving quite successful in the different Departments in which they were introduced. Mention was made of the very hearty cooperation of the men in all the departments affected by the Dilution, and this, Mr. Keachie pointed out, had proved of very material assistance in the successful introduction of the scheme.

The minute was in fact part of a 'stiff upper lip' response to an incident at Albion that had hit the national headlines in March that year and had been discussed in parliament at some length.

Willie Gallacher was an Albion shop steward with strong political views. Born in Paisley, the combination of a drunken father and a religious upbringing had made him a staunch teetotaller. Having been inclined to socialism from an early age, he had both joined and left the Independent Labour party by 1905, finding it lacking in Marxist principle. After working as a ship's steward for the Allan Line, he took a job with Babcock and Wilcox in 1910, but found the work too heavy. He next started work at Albion, operating a lathe. While there he joined the National Union of Brassfounders and Finishers, which he was to serve as an unpaid official during the war period. After being involved in the formation of the British Socialist Party in 1913, he went to the United States, where he absorbed the 'syndicalist' point of view. This school of thought distrusted both 'the bourgeois state' and the 'state socialist' parties within it. It believed in 'direct action', the political strike, using the trade unions to capture political power and administer the new society, especially its industries. By March 1914, Gallacher was again working in Albion, giving classes in economics and industrial history to fellow workers at lunch breaks. He was an opponent of war.

More important was the fact that he had become chairman of the Clyde Workers Committee. This group objected to several provisions of the Munitions Acts, including the prevention of employees resigning to work elsewhere without their employers' permission, and the dilution of labour. If this were to happen, held Gallacher, it should be under workers' control. A personal meeting with Lloyd George in Glasgow in late 1915 failed to settle the Committee's doubts.

On 17 March 1916, Gallacher precipitated a series of strikes on Clydeside. He insisted that he should be entitled to leave his work station and be allowed to investigate what was being done with unskilled labour, interview the women, examine their proficiency, rates of wages and other matters. When denied the opportunity to do this, he refused to take up the offer of a discussion with the local Commissioners responsible for the Munitions Acts, and called a strike, bringing out about 1,000 workers. Similar strikes were called at Beardmores, four more taking place there and elsewhere. On 28 March, the matter was extensively discussed in parliament, and while there was some support for Gallacher and his Committee from a Mr. Outhwait, most of the house supported the sanctions that had already been initiated against it. Two fellow members, John Muir and Walter Bell, were arrested with Gallacher. The committee's journal, the *Worker*, was suppressed and Gallacher spent a year in Calton Jail in Edinburgh. The movement collapsed, but Gallacher would subsequently spend a further three months in jail, in 1919, for his part in a mass demonstration in Glasgow's George Square in pursuit of a 40-hour week. In 1935 he was elected as a Communist M.P. for West Fife and would prove a thorn in the flesh of a future wartime government. He lost his seat in 1951.

There were no further instances of wartime disruption at Albion. The incident does show, however, that there was trade union activity in Scotstoun by the war years, and that several forerunners of today's unions appear to have been present, including the Amalgamated Society of Engineers and the Brassfounders and Finishers. The unions' activities would involve recruitment, the collection of dues, and perhaps some consultation on piece work rates. The payment of bonus in proportion to vehicles produced had always been made to key managers ever since the Finnieston days and had lately been extended to hourly paid workers, although details have not survived.

It is easy to imagine the conflicts that must have existed in the minds of trades unionists and manual workers generally during the Great War. During the years 1908 to 1914, wage rates had not kept pace with prices while the wealthy classes were becoming richer, and during the war, shortages caused creeping inflation to erode wages further, while the owners of munitions businesses did very well. Over against this, a basic sense of decency and patriotism, reinforced by respect for serving colleagues, operated in an opposite direction. There were bound to be tensions, and they were bound from time to time to be expressed. The period up to and including the Great War was one where established authority had come under increasing criticism, as witnessed by the upper-class Suffragettes. A literate working class, beneficiaries of state education since 1872, could observe, articulate its views and organise itself. In the course of the war, trade union membership in Britain doubled, rising from 3,959,000 in 1913 to 8,023,000 in 1919.

At Albion, it does appear that patriotism and a sense of self-sacrifice prevailed. One piece of evidence relates to further plans by the Ministry of Munitions to better utilise manpower, under discussion in early 1918. Exact details are missing, but *Motor Traction* records that at a mass meeting in the Whiteinch Burgh Halls,

4.2 Women shell production workers, 1916.

called on 4 February 1918 to endorse the proposal of the Amalgamated Society of Engineers to reject the government's plans, the motion was 'completely overturned by an overwhelming majority' of Albion workers. *Motor Traction* had published the story to disabuse readers of the notion that the whole of Clydeside had become obstructive.

In common with other centres of the munitions industry, such as Sheffield, Barrow or Coventry, there had set in a serious decline in workers' morale by the latter years of the war, and the government was forced to treat these areas with care. A number of special measures had already been taken. A separate Ministry of Labour had been created and additional munitions workers were encompassed in unemployment insurance schemes. Food rationing and subsidies, exemption schemes from conscription and wage arbitration awards were set up in the shadow of growing working class disillusionment. The government had learned at the start to treat Clydeside with particular respect. As early as 1915 it had passed the Rent Restriction Act, to keep rents close to pre-war levels, a move prompted by a rent strike involving the Clyde Workers' Committee.

If further proof were required of the Albion workers' reasonableness, it was the fact that the number of women workers employed under the Dilution arrangements rose from 44 at the start of 1916 to 130 in January 1917, and climbed to 472 by the time of the Armistice.

As a social phenomenon, the Dilution Scheme had positive effects at Albion. Girls from middle-class areas such as Kelvinside worked side by side with those

4.3 Line up of Albion War Office vehicles, 1915.

from the industrial areas of Govan and Partick and learned to respect one another, to appreciate each other's point of view and to form friendships. Quite apart from being largely responsible for shell production at Albion and thus helping in an essential war task, women such as these furthered the case made by the Suffragettes, and women were granted the vote without opposition after the war.

It is hardly necessary to emphasise the importance of vehicular transport to a war effort that involved the mobilisation of five point seven million fighting men and the stores that supported them. There were constant troop movements and ammunition, food, clothing, temporary accommodation and medical supplies had to be replenished all along the western front, which stretched from the coast of Belgium down to Switzerland. The logistical process began at factories such as Scotstoun, however, and convoys of vehicles were leaving Albion for the south twice a week in all weather conditions, a gruelling task for the squad of drivers involved.

A management meeting minute reveals that at some point in 1916, T. B. Murray had visited France, to see the vehicles in action. John Francis Henderson had also visited the main transport depot at St. Omer, near Boulogne and close to the Belgian border, to liaise on spares. Murray advised the meeting that reports received 'on all hands' concerning the running of Albion vehicles 'were extremely gratifying and reflected great credit on all who were engaged in their production'. One or two modifications had, to his mind, seemed necessary. Larger steel rear wheels were required 'in view of the very abnormal conditions existing in France'. In addition it seemed to Murray that if stronger radius plates (for wheels?) together with the provision of throttles that could be controlled by both hand and foot were sanctioned by the authorities, the popularity and efficiency of the vehicles would be further enhanced. It would in due course be revealed that there were more Albions employed on the western front than any other kind of vehicle. They were, moreover, employed in every sphere of British operations, as ambulances, mobile workshops, dental surgeries and all other manner of more basic uses.

4.4 15 h.p. Albion ambulance, 1916.

A review of production at a subsequent meeting in January 1917 revealed the extent to which wartime output was exceeding peacetime deliveries. 1,454 vehicles were built in 1916 and 927 had been built in 1915. In contrast, the figures for 1913 and 1914 had been 554 and 591 respectively. Plans were in hand to increase production further, from 30 chassis per week to 42 per week, but in the short term, the throughput of spares for the war effort made an immediate increase impossible. An insight into the introduction of shell production was also provided. This would take place in the upper flat of the new western section of the concrete block. After machining, shells were brought down by an elevator on the building's exterior. They were then transported to a press where the noses were shaped after heating in gas and oil furnaces. Next, they were varnished, examined by government inspectors and then despatched. In addition to the above, the toolroom was busy making jigs and fixtures for other munitions plants.

The management meeting minutes also reveal that in 1917, 1,689 vehicles were produced, and attempts to raise weekly production from 36 to 38 chassis had been thwarted by the demand for spares and by the transfer of 25 engine fitters to aircraft factories to work on aircraft engines. At the beginning of 1918, the government changed the shell contract, which had been for 8 inch shells, to 6 inch bottle-nosed shells, which required a redesign of production facilities. During these latter stages of the war, prices and deliveries of materials had continued to get worse. Some key components had gone up by 15 per cent in a year, others by 50 per cent, and the demand for aircraft parts was interfering with component deliveries for vehicles.

4.5 A wartime Albion A10.

One year later, on 17 January 1919, the management meeting was advised that production of vehicles had increased to 1,843 over the period. Murray took the opportunity to review Albion's contribution to the war, which had ended on 11 November 1918 'with particular reference to the manner in which Albion vehicles had behaved under the very severe conditions on the various fronts'. Official but confidential information had been received that for reliability and low running and upkeep costs, Albion vehicles 'were head and shoulders above all other makes'. A big factor 'in obtaining such happy results being, in the opinion of the War Office Authorities, the Murray-Albion system of lubrication'. A total of 5,594 vehicles had been supplied, together with 73,892 shells, plus an unknown quantity of engines, spares and jigs and fixtures. A quantity of Murray-Albion lubricators had also been supplied for the Dragonfly aircraft, and had also given great satisfaction.

As must be clear from the above, this had involved a heroic management effort. The workforce had risen from 940 in 1914 to 2,107 in 1918, and within these totals, staff had risen from 196 to 400. Exact details of the organisational structure do not survive, but aside from the directors, 11 line managers attended the first quarterly 'heads of departments' meeting in 1915, while 13 attended the same meeting at the start of 1919. Of those attending the earlier meeting, David Keachie was in charge of production, A. Donaldson was head of the Order (Purchasing) Department, T. L. Webster was head draughtsman, George Pate was head of vehicle engineering, G. Dykes was responsible for bodywork, R. Thomson in charge of depots and W. McFarlane was Company Secretary. P. Gilchrist was Chief Cashier, as was noted, and three others, P. E. Wright, A. D. Clarke and H. G.

4.6 Wartime convoy en route via Biggar in 1915.

Thorpe, judging by their remarks, appear to have been sales managers, most likely having different peacetime territorial responsibilities. Among the management innovations of these years was the introduction of a works progress scheme and the setting up of a production department, in the hope that they would cut costs through the monitoring of piece part production.

The impression is given that Murray was standing back from day to day design matters as never before. He was awarded his Doctor of Science degree by Edinburgh University in July 1917, for a thesis on carburation, which suggests that for some years he had managed to find enough personal space to pursue his own interests. It is also known that it was his practice to take the time to go over the design ideas of subordinates during these years, praising or making constructive suggestions as appropriate.

As was noted earlier, the company became a public company on 1 January 1915, and at this point Nicol Paton Brown took a seat on the board, together with George M. Young, who was in charge of the London operation. Brown was 62 when he joined the board, and had extensive commercial experience as a company director. The board, strengthened by outsiders, was in a better position to stand back from day-to-day operations than it had ever been before, leaving this more and more to subordinates, whose responsibilities and experience had grown over the war years.

At the start of the war, Albion, in common with everyone else, assumed that it would be short, and in its early wartime advertising, invited orders for

4.7 Biggar residents inspect War Office vehicles, 1915.

delivery on the cessation of hostilities. As the war stretched out and news was received of the titanic but ultimately indecisive battles of Ypres, Verdun and the Somme, the directors' minds began to focus more and more on the post-war period. This was by no means unusual, and speculation on the future of the commercial vehicle industry was beginning to feature increasingly in the specialist press. Among the widespread concerns expressed were the effects of the returning millions of troops to the labour market. Albion had considered one aspect of this, resolving to take back those workers who had left for military service. The next concern was the possibility that the War Office would flood the domestic truck market with vehicles that were no longer required. It was of course impossible to forecast the outcome of this eventuality, since everything would depend on the volumes released, domestic demand conditions and timing. Nevertheless, the question worried the Albion board and the industry generally for several years.

During the war, it had become apparent throughout industry generally that massive benefits were being gained from the production of a small range of products. In 1916, the advantages of variety reduction at Albion were seen most clearly. That year, 1,451 A10 vehicles were made for the War Office, 99.8 per cent

4.8 Bassett Lowke model of Albion A10 war vehicle.

of all units produced. The board accordingly resolved to concentrate in the post-war period on two models: the hugely successful A10, 32 h.p. model, and the A16, which had been designed on the eve of the Great War as a replacement for the extremely popular 16 h.p. chassis, known as the A3, which had been the mainstay of Albion's business from its design in 1904 until 1914.

A review of the evolution of Albion's model policy is appropriate at this stage. During the years 1900–1903, it had produced two-cylinder, horizontally opposed engined vehicles. These generated 8 h.p. until 1902, when the power of the engine was raised to 10 h.p. by widening the engine's bore. In 1903, in accordance with industry practice and customer preference, the change was made to vertical two-cylinder engines, initially rated at 12 h.p., but then raised to 16 h.p. in 1904, again by enlarging the engine's bore. The 16 h.p. chassis, known as the A3, which like previous Albion chassis was capable of either pleasure or commercial use, had a chain-driven rear axle. It had proved to be hugely reliable and, as was noted, formed the mainstay of Albion's business up to 1914. In 1906, the 24/30 h.p. pleasure car was launched, but lacking in support from Lacre in the high-volume London market, only produced sales of 57 units and was allowed to peter out in 1913, the year that Albion decided it would no longer accept orders for motor

59

4.9 T. B. Murray in DSc academic dress, 1917.

cars. The 24/30 h.p. pleasure car had given Albion the opportunity to develop a four-cylinder engined chassis for commercial use, and this had resulted in the 32 h.p. chassis, the A10, which had proved its worth in the years up to and during the Great War. Recognising a progressive drift away from chain-driven vehicles, Albion had in 1911 started to sell a 15 h.p. model which had a shaft-driven live worm axle, but its production had been halted by the war, as had the 16 h.p. two cylinder A3. In addition to the above, Albion had in 1914 launched a 25 h.p. chassis aimed specifically at the charabanc and two ton truck market. Again, its progress had been halted by the war. The war had provided the perfect focus and opportunity for rationalisation. It had reinforced the benefits of producing a small number of models while at the same time removing the difficulties that would be encountered in the market place with the phasing out of a wider range: it had to go to make room for wartime production.

4.10 Albion A10 mobile workshop.

In a period of eighteen years, Albion had developed nine models rated at 8, 10, 12, 15, 16, 25, 24–30, 32 and 20 h.p. Superficially, it seems a large number for a specialist producer of commercial vehicles, but this was a time of rapid evolution in the industry which was to culminate in the near-universal adoption of the four-cylinder vertical engine/live rear axle format soon afterwards. In addition, it was an evolutionary phase, where models were not completely recast, but substantially carried over to each other major developments in engine or axle technology. In this respect a simpler future seemed to lie ahead, however.

Financially speaking, the war helped consolidate an already successful performance, building up a foundation that enabled the company to cope with the post-war trials that lay ahead. The ordinary share capital, which had increased from £23,000 in 1907 to £70,000 in 1913, was raised to £200,000 in 1915 when the company changed to public company status. Six per cent preference shares had stood at £20,000 in 1907, were raised to £70,000 and £200,000 by 1915. In line with these measures, fixed assets rose from £97,000 in 1914 to £130,000 in 1918.

Conversion to a public company had been urged in 1913 by Jackson Millar, a minority shareholder, but the founder-directors were against this step lest they lost their grip of control, justifying their reluctance on the basis that they had never experienced any problems with the raising of capital, which was indeed true. Furthermore, when the change was necessitated for wartime expansion, the

4.11 An A10 in Ankara, recently photographed — a survivor of Gallipoli?

external capital was taken in by means of the issue of the preference shares, the equal balance between preference and ordinary capital being kept, as ever, by converting reserves to ordinary shares in a bonus issue. This left the owner directors in possession of the vast majority of the votes as a consequence of their still having the bulk of the ordinary shares. It could be argued with some justification, though, that this was a policy that denied Albion a far larger market share of vehicle sales in its earlier years. It was possible to creep along, grow by degrees and retain control, but Hozier Engineering, in particular, prior to its disastrous transfer to Alexandria as Argyll, had grown much faster than Albion in the years up to 1906. It would appear that to Murray, Henderson and the two Fultons, control was always more important than size.

Profits during the war years are stated in the accounts at £68,411 for 1915, £67,265 for 1916, £73,660 for 1917 and £96,466 for 1918, which gave an average return of 38 per cent on issued capital. These very large rewards were the subject of special wartime taxation arrangements. As a result of the need to increase revenues for wartime expenditure and the desire to exercise even-handedness in the raising of taxes, the government in 1915 imposed the Excess Profits Duty. This averaged 63 per cent of excess profits made over the wartime period. Details of its calculation in the case of Albion have not survived, but it

4.12 Albion A10 line up at Scotstoun.

seems to have hurt, as Murray is recorded as being angry at its continuation after the war was concluded. It was only repealed in 1921.

As far as wartime liquidity was concerned, Albion's balance sheets reveal debtors rising from £43,000 in 1914 to £219,000 in 1918, but with the debtors being in effect the government, there was no question of a liquidity crisis. Creditors rose correspondingly in these years, and stock by a multiple of three, but none of this could be construed in any shape or form as a liquidity problem. A bank overdraft of £27,000 had arisen by 1918, but was offset by a cash balance of £39,000 and investments of £29,000.

Regrettably, the details underlying these very satisfactory figures have gone, and particulars of sales values, margins and costings with them. It is likely though that much of the wartime business was done on a 'cost-plus' basis, and that any over-costings would be caught by the excess profits tax.

On 14 November 1918, three days after the Armistice, all of Albion's military contracts were cancelled. Notwithstanding the end to guaranteed profits and the uncertain future that were formally ushered in by this development, the company breathed a huge sigh of relief which marked the end of its involvement in the humanitarian disaster that was the Great War.

In due course, a memorial to those Albion employees who gave their lives in active service was erected outside the entrance at Scotstoun.

CHAPTER FIVE

Back to Peace: Boom, Bust and Buses

As the pace of normal life began to resume, the motor industry as a whole seemed well placed to benefit from several general trends that were already apparent before the war. Prior to 1914, the motor car had begun to be adopted by an increasing proportion of the middle classes as a serious alternative to horse-drawn modes of conveyance and the motor bus was becoming a more popular choice for local authority transport enterprises. Motor buses, mostly imported from the continent at first, had begun to be seen in small numbers from the turn of the century onwards and gradually came to be used as feeders to railways and on rural routes. Initially, the punishing routine of constant starting and stopping in large towns and cities was beyond the capability of the early models, but reliability was gradually improved and the carrying capacity relative to both passengers and goods had greatly increased by 1914. In London, horse-drawn buses were completely superseded by motor buses by that year, when it is recorded that there were 51,000 buses, coaches and taxis licensed to operate on the roads of Britain. The figure had stood at only 24,000 in 1910. By 1916 there were 331 bus operators in Britain. A growing market was there to be exploited.

Between 1910 and 1914, the number of licensed goods vehicles had risen from 30,000 units to 82,000 units. A high proportion of these were operated by traders and manufacturers for the conveyance of their own goods. The independent hauliers had yet to make their mark, but this was to happen progressively during the 1920s. The omens for commercial vehicles appeared promising. Local authorities were beginning to recognise an obligation to spend more money on road improvement, and earlier, Tarmac had been invented by combining tar and crushed slag. At last the clouds of dust that had bedevilled both motorists and pedestrians could be conquered, and better roads, then as now, encouraged greater levels of vehicle ownership and usage. But there was to be turbulence before the above factors bore fruit.

As soon as the war was over, the pent up demand for vehicles to replace worn-out and out-of-date fleets produced boom conditions at Albion, and the 1919 output of 484 20 h.p. chassis and 1,219 32 h.p. chassis was comparable with the best of the wartime years. An advertisement of 29 January 1919 confirms that the company was exploiting its wartime experience to the full. It read:

5.1 An A10 Albion leads the first post-war Glasgow Fair Celebration, Stranraer, on 19 July 1919.

Built on a
Long Life Basis

For over 18 years every Albion has been designed and built on a 'long life basis'. That is why you may today see Albions on the road that have given 12 to 14 years' satisfactory service. That is why the thousands of Albions supplied to His Majesty's Forces during the War are unsurpassed in their record for low cost of repairs by any other make. Built into an Albion is every feature that will pay for itself during the life of the Vehicle, in low maintenance cost and length of life...Albions are built by the largest manufacturers of Motor Vans in the British Empire.

The handsome output figures and high demand of 1919 had led to the largest profit figure achieved to date at Albion, £147,506. In keeping with these buoyant conditions, it was seen as time to expand, and a large new building was erected to the west of the office block. This came to be known, unsurprisingly, as the West Building. The time had come, too, to take action to expand the repair workshops capacity in London and Manchester, and ground was acquired at Willesden and Old Trafford. A Birmingham office was also opened. The finance for these developments came from two sources. Authorised ordinary share capital was raised to £400,000, and by the end of 1919, issued ordinary shares stood at £260,806. A further £189,400 was issued in six per cent debentures. This meant that, of the total capital issued of £584,000 by 1919, 45 per cent was in ordinary shares with the balance in fixed coupon stock. Lest it be thought that the directors had altered their views on control of the company, only £40,000 was brought in from the sale of ordinary shares in cash, the rest being issued to holders of ordinaries in proportion to their holdings, as a bonus issue. The directors' grip on control therefore remained as firm as before.

Profits for 1919 would have been even higher had it not been for a nationwide moulders' strike which affected supplies of cylinder castings from September 1919 to January 1920. This was a symptom of the general move leftwards across Britain that had been accelerated by the war. The number of strikes occurring in British industry increased dramatically during the years 1919–1921, and trade union membership rose to its apogee of 8.3 million in 1920. Apart from the knock-on effects of the moulders' strike, Albion was affected in several other respects by this tendency. The company, in concert with other major employers at this time, dropped the working week from 54 to 47 hours at the beginning of 1919, which necessitated a reorganisation of both the day shift and night shift to keep output up at the same level as before. At around the same time, several of the trades unions represented at Albion were involved in amalgamations, resulting in the presence from the early 1920s onwards, of the Amalgamated Engineering Union and the Transport and General Workers Union.

Over the period 1919/1920, two point nine million men in Britain were returned to civilian employment and half a million women dropped out of the work force. The discontinuation of munitions work at Albion was followed soon afterwards by the release of the female production workers and, in response to government exhortation, the company, in common with other employers, took on a 'considerable number' of workers disabled by the war. If the most lasting achievement of the trades unions during these years was the substantial reduction in working hours gained, then one of its negative effects was the enforced layoff of workers whose employment had depended on receiving parts from suppliers who were the subject of strikes.

By the turn of 1920, this had led to the suspension of a large part of the work force. By March that year, pent up demand had been satisfied and orders for chassis began to fall away. The night shift was stopped in October. It had run continuously for ten years, barring weekends and holidays. Next, the much dreaded and anticipated reduction in military truck stocks occurred. Of the 60,000-odd vehicles that had been employed by the forces, some 20,000 were put on the market by the Ministry of Munitions, at very low prices. This also affected overseas business, since military vehicles were also being sold in export markets.

The Albion response was logical. The company purchased back 1,000 chassis of the famous A10 type, some which were almost new, and some of which had given two or three years' service on the Western Front. A number of these machines could be sold by Albion as they were, but others were overhauled or rebuilt at a special shop temporarily occupied for the purpose at nearby Whiteinch. The vehicles were sold with Albion's six month guarantee. As a result of shrewd

5.2 Thomas Hamilton of Fauldhouse's 'demobbed' A10 military waggon, in use for haulage c 1920.

trading, the reluctant payoff of 350 workers in September 1920 and careful monitoring of the refurbishment costs of the ex-military stock, a profit of £104,223 was struck, remarkable for a year that had started well but fallen away rapidly. It was perhaps his frustration at the looming downturn that had caused Murray to rail, in May 1920, at the continuation of excess profits tax at the wartime level of 60 per cent. He considered this 'thunderbolt' a gross disincentive.

It is appropriate at this stage to pass from the study of Albion's business and focus on T. B. Murray's deep involvement in professional affairs during these years. His apparent withdrawal from detailed vehicle design work by this time together with the completion of his DSc and the appointment of competent subordinates allowed him to devote his attention to the affairs of the Institution of Engineers and Shipbuilders in Scotland. In 1919 and again in 1920 he was President of the Institution, which met in Elmbank Crescent, Glasgow. His inaugural speech of 1919 remains as evidence of his personality, engineering interests and width of vision.

Naturally enough, he began his comments with a reference to the war, which he was thankful had at last ended with 'a great victory and the nation at peace'. Murray considered it an honour that he was the first president of the Institution to have come from the motor industry, a symbol both of the emergence of motor engineering as a recognised branch of the profession and the ever increasing importance to mankind of motorised transport. It was of course logical that he

5.3 1923 24 h.p. military waggon.

should devote his speech to his own field of endeavour, and he began with some facts and figures which encapsulated the achievement of vehicle designers over the previous quarter century. He noted that in 1894, the car which won the Paris/ Rouen race weighed about 1,300 lbs. and developed three and a half horse-power, a ratio of 370 lbs. per one horse-power. By 1919, an average motor car only required 80 lbs. of weight per one horse-power, and he anticipated that shortly a ratio of 56 lbs. of weight per horse-power would be all that was needed. Already there were vehicles that undercut these norms.

As for engine design itself, tremendous strides had been taken forward with the development of the four-cylinder version, and six, eight and twelve cylinder models could even be found. Carburetters and magnetos were a 'marvellous achievement', and high speed engines of 2,000 or even 3,000 revolutions were able to develop high levels of torque, a far cry from the much slower early models. All this had been accomplished with increasing levels of fuel economy. There was room for improvement, however, and this had to continue, but, he concluded, 'the ideal engine which doubtless will be ultimately evolved may quite probably be an entire departure from the internal combustion type as it is known today'.

Murray next turned to the subject of vehicle speed, noting that the average speed of the winning car in the Paris/Rouen race had been 12 m.p.h. In the Grand

5.4 T. Blackwood Murray, late 1920s.

Prix of 1914, over a difficult course of 466 miles, an average of 65 miles an hour was attained. In the spring of 1919, a speed of 150 miles per hour had been achieved by a car on a motor racing track. For high speeds to be available to the general public, which he felt was inevitable, improvements in highways were necessary. As Murray put it,

> In the short space of a quarter of a century the safe speed of passenger transit on ordinary roads has been practically quadrupled. In future, when we have, as we no doubt will have, special motor-vehicle highways with recognised up and down tracks probably running over or under bridges to eliminate the crossing dangers of other intersecting highways, the speed of motor vehicles on these roads will at least equal that of express trains today, and with perfect safety.

In future, the road engineer would have to build roads of a breadth proportionate to the volume of traffic, perfectly smooth, dust-free and suited to pneumatic rubber tyres only. All animals would be prohibited, and traction engines or agricultural machinery would also have to be banned, so as not to destroy the road surface. Sharp bends and steep gradients would require to be avoided. Private vehicle owners would, of course, have legal redress for damages against authorities failing to maintain roads in good order.

The thrust of Murray's address was to survey the array of technological improvements that had enabled the motor industry to move so far in such a short space of time. It was to these innovations that civilisation in general and engineers in particular would have to look in the future to maintain progress. The first important technological improvement to be described by Murray was in the area of metallurgy. Over the quarter century, an almost total dependence by motor engineers on mild steel had been broken as alloy steels were developed. Using these, component sizes could be reduced by as much as two thirds. Stresses of up to 30 tons per square inch could be sustained by the new nickel chrome steel, for example. As a result, the dimensions of gear wheels, in particular, were far smaller than could be achieved in the past. Murray noted that not all of these improvements were the result of superior materials. Many resulted from the development of heat treatment and related techniques to harden metals. For this reason the heat treatment departments of motor works were in the charge of a 'technical expert', who fully understood the processes of normalising, carburising, tempering and reheating.

Murray next broached the subject of machine tools, a field where immense strides had also been taken over the period. Here, improvements had resulted in 'the evolution of tools for the production of parts in quantity in a cheap and efficient manner'. One of the most interesting developments was the advance made in automatic grinding machinery. This had enabled complicated parts such as multiple-throw crankshafts to be produced 'with an accuracy unapproachable even by a skilled turner', and at a far lower cost than previously possible. Automatic grinding machines had also enjoyed considerable success in the rectification of the teeth of hardened gear wheels, which became slightly distorted as a result of heat treatment. Broaching machines were also being used for increasing numbers of applications, such as the machining of the castellated holes in the bosses of wheel parts.

In conjunction with developments in machine tools, high speed steels had also been introduced for use on cutting tools, enabling cutting speeds to be substantially increased. This had in fact led to a need to completely redesign the machine tools. Much stiffer and more robust designs were needed to cater for the higher cutting speeds.

In future, Murray foresaw the need for the elimination of jigs through the machine tools themselves becoming live jigs, 'designed and constructed to carry out special operations on the particular part and nothing else'. Already, multiple-head milling machines were available that would surface three or four faces on a cylinder block at one setting. Machines were also available that could simultaneously drill upwards of 50 holes on a cylinder block on a number of different surfaces. This indicated the direction in which evolution was tending,

5.5 A10 lorry supplied to South African Railways in 1918 (photographed in 1932).

and 'wherever the quantities to be produced warrant the expense, it is the correct policy to adopt'.

It hardly needs to be said that most of Murray's predictions came true. Modern motorways emerged in the 1960s in Britain almost exactly along the lines predicted, and the Wankel engine is only one example of a radical approach to internal combustion design that could yet pay dividends and fulfil Murray's prophecy regarding engines. Metallurgy and machine tools continued along the lines of evolution he had highlighted, leading to the transfer machine and eventually, into robotics. It is impossible to say to what extent Murray was merely passing on second hand the observations and predictions of others. That is not, perhaps, the point. Whatever his sources might or might not have been, his paper reveals a mind capable of dealing with the smallest detail and yet not neglecting to look ahead to the larger picture. Murray's talents in this direction and his initial vision regarding the future of the motor car must surely, at least in part, explain the emergence of Albion at the end of 1899. It was important, too, that such an imaginative mind sat at the head of the company's board.

But how was it possible that such a visionary speech could be uttered by the chairman of a company whose technological conservatism and love of slow evolution had taken it, early, into commercial vehicles? The explanation might be simple: an ability to predict is not necessarily the same as an ability to innovate. Alternatively, it is possible that the early constraints on the recruitment of directoral level manpower, linked to the board's desire to retain control, forced Murray to specialise. It is equally possible, in the absence of firm evidence, that a younger Murray simply liked to do it all himself, which would alone explain the

5.6 War Office subsidy chassis outside Alexander McAra's garage, Dundee c 1923.

limited range of technical development the company undertook. A final conclusion cannot now be drawn.

The years 1921 to 1923 saw Britain still in the grip of a trade depression. Industrial production and manufacturing output generally had fallen during these years, after the immediate post-war boom. These conditions applied throughout many parts of the world. The coal mining industry, still benefiting from a seller's market, suffered a three-month long strike in 1921. Albion, in common with many other manufacturing businesses, was forced to close down intermittently that year as fuel supplies fell. 1922 saw the depression continue on a worldwide basis, affecting overseas sales, and the War Department continued to release surplus military vehicles into the home market at rock-bottom prices. Things began to improve in 1923, randomly in the first six months of the year, and steadily in the latter part. Towards the end of that year, the Australian market began to pick up. In 1921, Albion had formed the Albion Motor Car Company, (Australasia) Ltd., and by 1923 orders, especially from Sydney and Melbourne, were growing substantially under the new selling arrangements.

Statistics for these years show works employees standing at 497 in 1921, 527 in 1922 and 620 in 1923. Staff levels for these years were 193, 248 and 220. Profits halved to £50,792 in 1921, fell to £10,945 in 1922 and rose to £46,157 in 1923.

5.7 General view of McAra's, Dundee with Albions outside, c 1923.

The output of vehicles for 1921 was only 166, rising to 244 in 1922 and 606 in 1923 as trade picked up. During these depressed years, however, a great deal of building for the future took place. The new Manchester repair shop was opened on 17 January 1921, and the Willesden repair depot on 24 May that year. The branch office that had opened in Birmingham was closed and replaced by a service and repair depot at Cox St., St. Paul's Square, Birmingham, which was opened on 29 November. In Australia, the improving business conditions led to the takeover of larger premises at 103a Madeline St, Carlton, Melbourne, which could store vehicles, carry spares stocks and had the capability to undertake repairs.

As far as model development was concerned, it had been recognised in 1920 that the A16, four-cylinder, 20 h.p. vehicle had been designed with too slow a speed compared with the competition in the market place, and it was replaced with the A20, which had first been revealed at the Olympia Motor Show in October that year. The A10 32 h.p. vehicle which had proved so successful in wartime was brought up to date and launched in October 1923, as the A10 Mark II. Thus, the two ranges on which Albion had decided to specialise in the post-war period were updated.

Trading conditions nationwide and worldwide were improving by 1924. Large orders were received from the India office at the end of 1923 and the U.K. War

5.8 Two Albions, one Maudslay and one Ford van at Rowe Brothers, Bristol, 1920s.

Department also placed a large order for a special type of 24 h.p., 30 cwt. chassis. This type of vehicle carried with it a subsidy to civilian users, and for this reason Albion decided to bring standard vehicles into line with the War Department's requirements by fitting a 24 h.p. engine to 25, 30 and 40 cwt. models. The decision to standardise production by producing only two models was therefore beginning to prove unfeasible in the light of market conditions. The model situation was also beginning to be affected by developments in the market for buses.

As well as the general growth in the bus market that had occurred prior to the war, there were special developments that could not have been predicted. Motor buses had been commandeered by the military in large numbers during the war, which caused a general shortage of buses at home. Petrol supply shortages combined with this to cause many bus operators to curtail their wartime operations. The only exceptions were those operators, such as the Midland Red service, whose petrol-electric buses were not required by the War Office. These firms were able to consolidate their trading position. Because of the increasing use of buses on the western front as the war progressed, a large number of men came to learn how to drive and run these vehicles. When they demobilised from the armed forces, the release of buses onto the market place, at very low prices, caused many of them to enter the bus business for the first time. Some used their war gratuity to purchase one vehicle and start a family business. Bus operations therefore expanded in unforeseen ways. The history of the bus business from this point onwards, if not characterised by uniform progress, was on the whole one of general growth. Buses licensed in 1919 stood at 44,081. By 1923 the figure was

5.9 1923 Albion photographed at Bombay docks in 1958.

85,956, although the official totals include taxicabs at this stage, perhaps in the region of 50 per cent.

The bus business seems to have quickly fallen at this time into four main divisions: the provincial omnibus company of stability and substance, the small company in family hands, said to be careless of good will and keen to make quick profits, the owner working in an area established by a larger company simply to encourage a buy-out offer, and 'pirates' who ran ahead of scheduled services. Only the stable operators ran regular services, and it has been estimated that in 1922 only 200 operators out of 1,888 fell into this category.

The ease or difficulty of entering the business varied quite arbitrarily from one centre to another. There were urban or semi-urban services at regular intervals, some worked with the agreement of local authorities. Rural services ran from a few miles up to about 30. Most main and secondary roads acquired services throughout the decade. Long distance services lay ahead and pneumatic tyres were fast becoming the preferred type for bus operations.

It has to be observed at this stage that Albion's interpretation of trends and response to the above was less than wholly visionary. Before the war, chassis suitable for charabancs were being built as A10, 32 h.p. vehicles, with a maximum carrying capacity of 29 people. A 25 h.p. model had been developed before the war for both truck use and 25 seater charabancs, but it had been dropped during the hostilities. Special 'torpedo' style, part aerodynamic bodywork had been developed for this vehicle, which could be had with or without hood, and it was

75

5.10 18 passenger Viking coach, 1923.

regarded as ahead of its time. A management meeting of 12 September 1921 notes a belief at Albion that road sizes would restrict bus dimensions and carrying capacity for the medium term, which with hindsight was not what happened. In spite of having launched the new Albion Viking coach at Olympia in November 1923, the company was not in a position to take advantage of the rapidly improving bus market. The Viking's sleek lines and tapered nose, which gave it the appearance of an expanded passenger car, only extended to the carriage of 20 passengers. Adjustments would be required.

The question of models was one which the directors commonly addressed during these years. At one stage in 1922, the possibility of producing only one model had been raised, but rejected because it was felt that demand would not be large enough to keep the factory up to capacity. The initial plan for two main models was itself beginning to come under pressure from outside forces, and clearly all was not well on the bus side of the business. The rigid policy would have to go.

The outcome of this response to external stimuli was a rapid widening of the vehicle range between the middle and end of the decade. In 1925, an 'overtype' A10 3 tonner was produced. In this type of vehicle, the controls were set forward of the engine, improving driver visibility and introducing a format that would reappear time and time again for many years to come. A new 29 seater bus, the Model 26, was launched at Olympia in 1925, and made a non-stop run from Glasgow to London and back in 24 hours five minutes. In 1926, the A10 32 h.p. chain drive vehicle was finally replaced with a live axle model, the Model 27. The same year, a new overtype passenger model (Model 28) was launched, capable of

5.11 First non-stop London to Glasgow bus, Model 26, 1925.

carrying 32 people. An improved Viking coach was also introduced. In 1927 a new 4-ton overtype chassis, the Model 35, was introduced, followed by a bonneted version, the Model 34. A five-ton version of the Model 35 was also made available. These were the first Albions designed for heavy work with the overtype layout, which gave better access to engines for repair and maintenance, a larger platform area than before, a short wheelbase and a small turning circle. In 1928, a new high speed chassis of 30 cwt., the Model 40, was produced, with an export variant, the Model 41. The passenger models previously designed were also offered in six-cylinder version (notably, the Viking Six).

In the absence of detailed records, it is difficult to know what caused so many rapid model developments across the whole range. Whether demand from the market forced the proliferation, or whether it stemmed from an attempt by the directors to anticipate market demand, it is difficult now to tell. The changes in the van sector of the models can be explained clearly enough. The Model 20, successor to the A16, was too expensive and large for the 15–20 cwt. sector of the market, and it had been discovered that a Dennis 30 cwt. vehicle could be bought for a massive £180 below the Albion equivalent, which admittedly, could carry a larger payload. In addition, the Albion directors felt that their vehicles in this range were outdated, and hence new designs had been prepared. It is likely that some of the changes, too, came from observing the movements of competitors, in an industry where it is usually difficult to say who is 'first' with anything.

Another subtle element in the mix of possible explanatory factors for the many model changes relates to the aesthetic dimension and attempts that might have

5.12 Albion Model 27, 3 ton tanker in livery of Shell, Melbourne, 1926.

been made, through engineering, to alter the outward form of vehicles. As early as 1920, a management meeting minute had recorded the following:

> Mr. Watt raised the question of appearance as a Sales factor. The general appearance of the vehicle, he thought, had a considerable value from the Salesman's point of view, and it was for this reason that he would like to get an expression of opinion by the Salesmen on the matter. Mr. Watt compared several of our Competitors' vehicles from an appearance point of view, and held that vehicles of the [War Department] subsidy type had taken a firm hold on the public, largely due to their massive appearance and suggestion of great power and durability. The general opinion was that in designing the new... model, something along the lines of the subsidy model should be aimed at.

Factors such as these could clearly no longer be ignored.

For similar reasons, Albion's Mr. Thomson had during the last year of the war suggested an improvement to the small scroll design name plate that decorated the radiator of the A10 'from an advertising point of view'. It was practically impossible to read the name from a passing vehicle. Thomson felt that the word

5.13 B.P. overtype tanker, late 1920s.

5.14 Carter Paterson overtype truck, late 1920s.

5.15 Albion 20 seater bus at Scotstoun, late 1920s.

5.16 Albion bus in Redland livery, late 1920s.

5.17 20 h.p. tractor and trailer, 1920s.

'Albion' cast in the top part of the radiator in clear black letters would be 'a much better arrangement'. The meeting at which this was suggested gave its approval, and members of the management were invited to submit sketches to Murray. The net result was the series of distinctive, pedimented radiator tops that featured much bolder designs, one consisting of the word 'Albion' in large, cursive Art Nouveau script. Sometimes, in the middle 1920s, this would appear together with the distinctive 'sunrise' motif that became Albion's hallmark, either taking up the whole radiator top, or placed on a plaque on the middle section of the top. In due course the sunrise-based motif became dominant, with the majority of vehicles coming to sport the larger, bolder version for many decades to come. The reasons for the gradual appearance of the radiator emblem in this manner have now been lost, but as with other developments in the 1920s, it is perhaps safest to say they were evolutionary.

This is a term that could only be applied to a certain extent to the engineering that underpinned Albion vehicles during the 1920s. The development of low-frame and semi-dropped frames for the buses produced in 1925, although possibly derived from AEC practice, was quite radical rather than evolutionary, as was the provision of six-cylinder engined options. At the same time, live axles had gradually succeeded chain axles, and pneumatic tyres continued to succeed solids as engines, gearboxes and axles at Albion were all gradually improved. The

81

5.18 Albion Model 40 bus owned by City of Oxford Motor Services, 1929.

company's engineering approach at this time is well summed up in a 1929 report on an Albion Viking bus, taken from the magazine *Coach and Bus*:

> Possibly because the Albion factory is geographically well away from the more usual centres of vehicle chassis production, the machines emanating from it have always displayed a certain amount of originality. This is not to infer that they are either freakish or even particularly unconventional, but rather that there is never any slavish adherence to established methods of construction or detail design if a sound or valid reason for an individual means of achieving the desired end presents itself.
>
> Thus the name Albion has always been associated with distinctive chassis, and coupled with it has been earned a reputation for reliability acquired in the days when the Albion was a heavy goods-carrier with a very deliberately governed two-cylinder engine. That type, restricted in speed, was necessarily reliable, given good general design and excellent material, and its subsequent longevity, while being an excellent advertisement (which it still is), may have become something of a mixed blessing to a works that still has to supply not merely pre-War but almost pre-motor era spares.

As examples of the above, the reviewer noted that a 'distinctive Albion feature is the heating of the induction manifold by exhaust gas', going on to observe that

5.19 Promotional drawing of a Model 28 Viking Six in City of Oxford livery, 1928.

'A method of engine mounting peculiar to Albion Chassis is adopted'. The transmission 'embodies a number of detail departures from the majority of contemporary chassis' and the worm-driven back axle also had 'unusual and yet very practical and desirable features'. He speculated that 'it is possible that the shipbuilding influence of Clydeside is more apparent than the automobile influence of the Midlands in some of the detail work', concluding that

> the Albion may be summed up as a thorough engineer's job, and by engineer I mean the man who deals with the machine in all the varied problems of actual use rather than the man whose mechanical ideals put him in touch with those who hold the steering wheel and wield the spanner.

The net market effect of the engineering design and model policies pursued by Albion from 1924–1929 was a very positive one, especially overseas. Generally speaking, Britain's exports throughout the 1920s had been steadily declining, especially in the area of manufactured goods, but the company was part of a growth sector which defied the trend. This was reflected in large orders for the Indian Government and various other 'Colonial Governments'. During 1929, again, large Indian orders came in, this time for 300 six-wheel chassis, which had only been produced since 1927. The detailed statistics no longer exist, but it has

5.20 20/36 h.p. school ambulance for Dumbarton Education Authority, 1929.

5.21 Lorry in Scottish Co-op livery, 1920.

been estimated that up to a fifth of Albion's output at this stage and beyond was being sent overseas, a percentage that would be sustained until the late 1930s.

In support of these overseas sales, Albion executives, as ever, conducted extensive tours. In 1924, T. B. Buchanan left for a six months tour of India, Ceylon, Straits Settlements, the Dutch East Indies and China in order to gain first hand information on the state of the markets there. R. Thomson visited Australia, New Zealand and South Africa from May to December in 1926, and visited India briefly again at the end of the year. G. M. Young left at about the same time for the Gold Coast and Nigeria.

As far as domestic sales were concerned, these too exceeded pre-1924 levels, with a widely spread range of orders extending across hundreds of customers from the government to the Co-operative Wholesale Society. Albion's main city depots, to which Bristol had been added in 1926, combined with local concessionaires to lift the company out of the sales depression of the post-war slump.

Some key statistics for the period summarise the progress made:

KEY ALBION STATISTICS, 1924–1929

	1924	1925	1926	1927	1928	1929
1) Vehicle Sales Units	1,100	1,309	1,295	1,489	1,658	1,538
2) Profits £000s	91	122	108	123	96	72
3) Issued Capital £000s	623	592	590	586	571	601
4) % Return 2)/3)	14.6	20.1	18.3	21.0	16.8	12.0
5) Nos employed: Works	950	1,099	1,213	1,361	1,733	1,254
Staff	310	454	493	493	601	607
Total	1,260	1,553	1,706	1,854	2,334	1,861

These years of relative certainty were punctuated by several memorable events. In December of 1924, the company celebrated its Semi-Jubilee. The company's concessionaires in Britain and Ireland were invited for a works tour which was followed by a dinner in the St. Enoch Hotel, Glasgow, attended by a total of 52 directors, heads of departments and guests.

The next notable event in Albion's calendar was the General Strike of 1926. This proved to be a watershed in British industrial life. The most militant of Britain's unionised workers, the miners, had demanded nationalisation of the pits just after the war and distrusted both coal owners and government. They had suffered wage cuts and worsening conditions from 1919 onwards, and had been forced back to work in June 1921, when their earlier strike was unsuccessful. In July 1925 the Trades Union Congress had pledged its full support for the miners, and this had been called up on 3 May 1926, shortly after a government subsidy supporting miners' wages, introduced to avoid confrontation, was withdrawn. The T.U.C. called on all its member unions to strike in support of the miners, and this lasted from 3 May to 13 May. Engineers and shipbuilders were called out on 11 May. The strike was virtually complete in every centre of membership. It was called off on 12 May, to the surprise and puzzlement of the strikers. The miners carried on striking for nearly eight months, but the carefully amassed stocks of coal, which was not in any case in great demand, resulted in the miners being starved back to work. Thereafter, general strikes were made illegal, although the appetite for these had been knocked out of the unions, and a period of relative industrial peace ensued.

The reversals experienced by the trade unions, at least at local level, are unlikely to have given Albion's directors any special pleasure, judging by a management minute of September 1921. When the company had been forced to shed labour at this time, they had kept shop stewards in employment. The minute noted that this had 'produced a fine spirit in the workers. Shop stewards now see that we consider them a help, and when this fact is established, it will be the means of getting the best type of workman to take this office'.

The third notable event of the period was the visit by the Prince of Wales to Albion's plant on 4 November 1927, the day of the opening of the Scottish Motor Show. He arrived at 12.30 p.m., accompanied by Lord Weir, Air Marshall Sir H. Trenchard and General Trotter, his A.D.C. T. B. Murray was absent as a result of a chronic illness that had begun to plague him several years earlier, and the Prince was received by N. O. Fulton and J. F. Henderson. These two presented the other directors to H.R.H. in the boardroom, at which point he asked whether or not Albion 'worked on mass production lines'. The Prince was informed that they could not, since the big range of models made this impossible.

Although the point was made lightly and without analysis, it must have raised a question in the minds of the Albion directors, whose departure from a one- or two-model policy that might have allowed mass production has already been described.

The 1920s were, for Albion, a period of managerial consolidation and change. On 17 June 1924, David Keachie, by this time a faithful servant of twenty five years, had been made a director, as were R. Thomson, J. D. Parkes and W. A. Donaldson. In 1929, A. E. Wright, the Scottish Sales Manager, was appointed a director. David Keachie retired at this point, as did Nicol Paton Brown. William Pate, BSc, was appointed director and works manager on 4 November 1929, and his brother George Pate, BSc, rejoined the firm as a director, having acted as Murray's assistant between 1910 and 1918. William Pate had worked in a range of engineering plants, and had latterly been managing director of Anderson-Grice Co. Ltd., manufacturers of gas engines, electric cranes and stone-cutting machinery at Carnoustie. Between 1918 and 1929, George Pate had been with motor manufacturers David Napier and Son Ltd. of Acton, leaving his position as a director to rejoin Albion.

The departure of Keachie and the arrival of the Pates was no routine reshuffle. The changes were precipitated by the tragic death of T. B. Murray on 11 June 1929 at his home at Monthey in Switzerland. Since about 1925/1926 Murray had been ill with consumption and had effectively been on leave of absence from the firm, from which he formally resigned in February 1929. He had received copies of board minutes and important management papers up until this point, and had

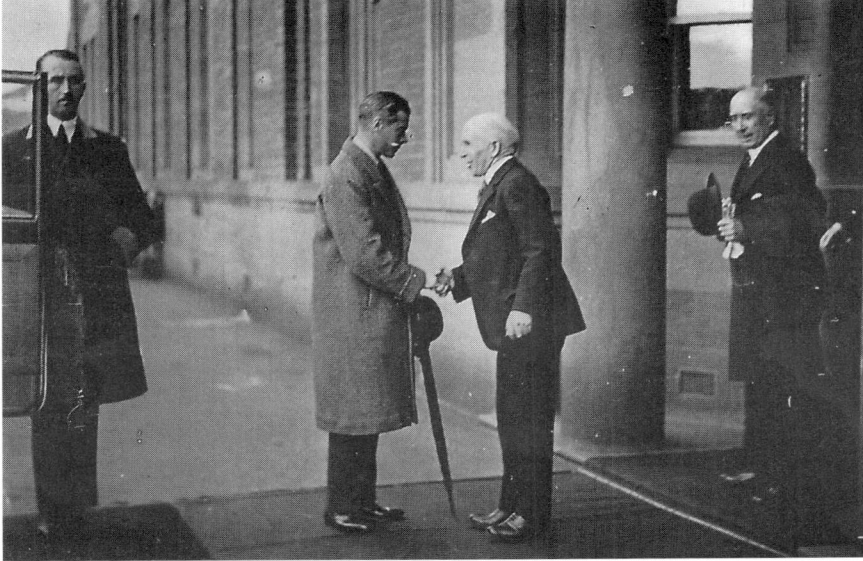

5.22 The Prince of Wales and N. O. Fulton at Scotstoun.

5.23 The Prince of Wales and N. O. Fulton with David Keachie at far right.

5.24 Fulton and the Prince of Wales examine machinery.

5.25 The Prince of Wales examines a dog cart and an A10 war waggon.

occasionally expressed his views in writing on matters that were being considered by the board.

By the beginning of September 1929, the board had decided to confront Keachie over the question of problems in the production areas, which they felt had been getting out of hand for several years, whereupon Keachie resigned. He was only two years away from retiral, and was given payment in lieu of notice and a lump sum approximately equivalent to the salary he would have earned to retirement. Records of the production problems do not survive, but the scant evidence suggests that Keachie may not have been coping with the increasing complexity of managing a wide product range.

It is known that Murray set great store by long service and loyalty, and it would therefore have been unnecessarily stressful to him for the board to take action against Keachie during his lifetime. N. O. Fulton had taken over from Murray as chairman (a role which, de facto, he was already fulfilling), with George Pate taking over the managing director's position. Murray's death had signalled the end of an era in more ways than one.

The Albion log book records his death in the following terms:

> Though only 58 years of age, he had lived a full life and has left behind him many happy memories of his Genial and Manly character, his unbounded keenness and energy, and his outstanding Capacity as an Engineer.

He was laid to rest on 19 June at Biggar Kirk alongside his forbears, his coffin having been borne up the hill from the railway station on a brand new Albion lorry.

CHAPTER SIX
Radiations

The most useful means now available for assessing something of the mood, interests, composition and character of the Albion workforce during these years is the works magazine that came out in January 1928, appropriately named the *Radiator*. The *Radiator* lasted until 1937, a full ten years. It was no propaganda organ, designed by management for the cynical manipulation of opinion to a one sided point of view, as many internal publications in other organisations have been before and since. It is doubtful if it could have been sustained so frequently and so long if that were the case. The *Radiator* was published monthly, with a gap in August each year to allow for the holidays. What is more, the employees paid for it, and for a modest fee could have each annual volume bound.

Its purpose was expressed in a note from the Albion board, printed on the first page of Volume 1, No. 1:

> For many years past we have felt the need of some medium whereby the community of interests which exists among us could find expression...There is an urgent call on employees, present or past, at home or overseas, to assist in making it a great success.
>
> We hope our magazine will long continue to radiate good will and good cheer, along with much interesting information on the many social and recreational activities in our midst, with occasionally a little useful knowledge thrown in.

This communication was signed by N. O. Fulton, who had been keen to get something started.

At an earlier stage, a short publication known as the *Compass* had served as a medium for carrying news of works clubs and activities, but had fallen through. This prior communication appears to have operated without the company's financial support, and seems to indicate a certain *esprit de corps*. In 1922, a full edition of the *Radiator* was compiled and made ready for printing, but never proceeded with. The reasons for its non-appearance have now gone, but it seems likely that the cutbacks and manpower reductions made at this time were on reflection considered an inappropriate context for the launch of an in-house magazine. The 28 January, 1928 edition was therefore the third attempt to set something permanent in motion.

The *Radiator's* first editor was William Lang, M.A., Albion's Welfare Supervisor, a former schoolmaster who had joined Albion from the educational world in 1919 at a stage where the company felt it needed such an appointment. Lang was a great lover of 'plain folks and plain things', and appears to have been a well liked figure at Albion. His main duty was the supervision of the firm's many apprentices, which at one stage had numbered in excess of 300. Lang was a good communicator, evident from a verbatim transcript of a speech he gave at a 1921 management meeting on the subject of intelligence and aptitude testing techniques for apprentice selection. Lang affirmed the social and recreational bias of the *Radiator* in his opening editorial, and attempted to encourage as many works and staff correspondents as possible.

The *Radiator* was originally eight pages long, but was soon extended as a result of its good reception and the volume of contributions. The first edition contained a black-and-white portrait and short biographical sketch of T. B. Murray, followed by an illustrated account of the visit of the Prince of Wales to the Albion factory. If this material gave the magazine a slightly paternalistic tone, it can perhaps be excused in a first edition for which employee contributions are likely to have been in short supply. The *Radiator* next featured communications from the Albion depots in London, Manchester, Sheffield and Leeds. The London depot reported on good business from the Olympia Motor Show and wished the *Radiator* well. Manchester provided a short history of the depot there, while Sheffield offered a list of 'things and people we would like to see and know', which included 'other people's opinions on six wheelers', 'A real summer for a change' and 'The Editor's face when he reads this rubbish'. Leeds sent a rather stiffly worded congratulatory note. Contributions from the depots became a regular and varied feature, and humour was a keynote, sometimes coming in the form of limericks (mostly very bad!) and sometimes in the form of cartoons.

The first edition of the *Radiator* contained a sketch and description of the new Albion Recreation Ground at Yoker, and it is possible that it was this development that triggered off the journal's appearance. It would have made sense to publicise the new facilities in this way and then use the magazine to consolidate and expand existing social activities around them. This would help sustain a happy workforce. To look upon this as manipulative would be to misunderstand the huge natural sincerity and kindness of Fulton. A sketch of the Yoker layout shows a nine-hole putting green, two football pitches, a pavilion, two bowling greens and six tennis courts. The grounds were bounded on the north by a stream that showed promise as a venue for fishing trips, the Yoker Burn. There was, in addition, a Club House. In describing these facilities, Lang observed that they were 'the most recent token of the sustained interest which our Directors have always taken in the wellbeing of everyone of us, from office boy or youngest apprentice, in our common task of maintaining the fair name of "ALBION".'

Another regular feature of the *Radiator* introduced in its first edition was its short accounts of 'Works Visits'. It would appear that apprentices were expected to attend, but the rapid fluctuations in numbers in both directions suggests that others came out of interest also. The first of these to be reported on was a trip to Ioco Rubber and Waterproofing Co. Ltd. of Anniesland, followed by an account of a trip to the Steel Company of Scotland's Blochairn

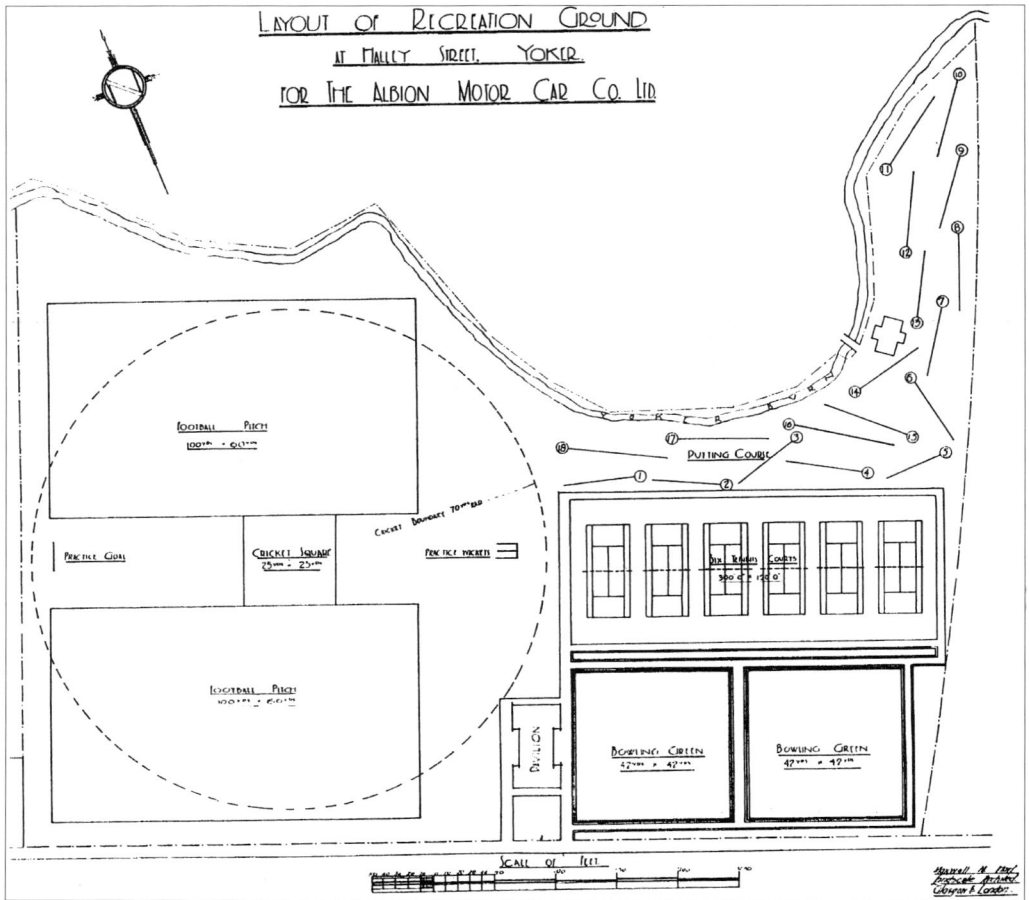

6.1 Layout sketch for Albion recreation ground, Yoker.

Works. Both trips were well attended and enabled visitors to study the production processes in full. Future visits were reported on at venues such as British Oxygen of Polmadie, G. & J. Weir of Cathcart, Yoker Power Station, Glasgow Corporation Refuse Power Works at Govan, Clydebridge Steel Works, and Glasgow Corporation's Transport Department's Coplawhill Car Works (for tramcars) and Larkfield Garage.

The first *Radiator* reveals that a number of clubs had been in existence in Albion for a long time. The oldest of these dated back to 1910. That year the Drawing Office Magazine Club was formed in the Drawing Office Extension, a galvanised iron structure that stood where the main block was subsequently erected. The members paid a small weekly subscription to buy magazines, which were shared on a rota basis. A more important club was the Albion Social Club. It was formed early in 1919 to promote social activity amongst the staff, which included outings, sports events, dances and so on. It ran golf and bowling tournaments, weekend rambles to beauty spots in and around Glasgow, concerts, dances, social evenings

6.2 Bowling Club Committee and Recreation Association officials (including Keachie and Mr & Mrs N. O. Fulton), 1928.

and whist drives. An annual concert was run to raise funds for the St. Dunstan National Institute for the Blind. Billiards tournaments were also run, and the Social Club in addition took responsibility for arranging presentations to staff when these were required.

The first club to feature in the *Radiator* was the Football Club. In 1922, a single Albion team had struggled to keep going in the midst of the trade difficulties of that period, taking its place in the Southern District Welfare League along with teams from Stephens, Yarrows, Clydebridge, Howdens, Dunsmuir and Jackson, Weirs, N. B. Loco, Alleys, Drysdales, P. & W. MacLellan, Howdens 'B' Department and Dalmuir. By 1928, there was both a Junior and a Juvenile team, and G. Reddy, who had been team manager in 1922, was president of the club. The Juveniles were described in the *Radiator* as being 'in every way true sportsmen', who were 'always ready to be the first to forget any incidents which may happen on the field'. By the late 1920s the club and its two teams were prospering both on and off the pitch. It occasionally organised seven-a-side tournaments within Albion, between departments. Home games at the new Yoker facility were played, where possible, on alternate pitches so as to save the playing surface. Football's popularity in the west of Scotland and its own internal momentum would ensure that the Albion Football Club would remain a permanent feature of life.

The availability of the new facilities at Yoker both strengthened existing clubs and brought new ones into existence. A bowling club that had 92 members in 1924 had 144 by 1928. It participated in an annual fixture with the Beardmore (Paisley) Welfare Club which consisted of a game at each company's premises, and as well as having bowling as its central activity, it organised whist drives and dances. The culmination of the year was the Annual Social and Presentation of Prizes, which in 1928 was held in the Trades House Restaurant, Glasgow.

Another club benefiting from the opening of Yoker was the Tennis Club. It had been in existence before 1927, playing its games on the grass courts in Scotstoun Showground. On 6 August 1928, the new courts at Yoker were broken in, the *Radiator* noting that

> Saturday, 6th August, was a red letter day for all Albion tennis enthusiasts, when the new courts at the Recreation Ground at Yoker were opened. An American tournament was held under ideal conditions, and was enjoyed to the full both by players and spectators…The ladies of the club deserve every credit for the way in which they catered for the hungry and thirsty crowds, and their efforts were much appreciated.

Clubs which did not require the use of Yoker had already existed for some time, and continued to prosper. An important member of this group was the Swimming Club, which started out as a ladies' club, but by 1928 had a men's section. The club based itself at the corporation baths at Whiteinch, meeting on a Thursday evening. Swimming instructors were found in 1928, and the club sessions consisted of a mixture of instruction in swimming and diving, water polo and occasionally, races.

The completion of a cricket pitch at Yoker and the delivery of equipment soon enabled a cricket club to begin. An initial membership of about 50 in 1928 allowed a team to be formed, and before long, it was involved in evening and weekend matches with other clubs. The activities of a golf club are also described in subsequent editions of the *Radiator*, and at one stage a shooting club also existed. In due course a motor cycle and car club was formed and this organised outings to such places as the western highlands around Arrochar. These consisted of cycle scrambling (on a competitive basis) and sometimes speed trials. Prizes were awarded and the social side rounded out through the provision of picnic lunches.

As the Yoker facility became more heavily utilised, an overarching Recreation Club was formed to which other clubs could become affiliated. One of its obvious aims was to co-ordinate the use of the new grounds, and board members took their place among its office bearers. The Recreation Club also took over responsibility for the maintenance of the grounds, which involved the appointment of a groundsman, the costs being met by a small voluntary weekly levy from employees.

Quite apart from the more overtly social activities at Albion, there also existed a Mutual Aid society, where contributors were able to rely on financial support when employees or their dependents fell on hard times, as in cases where serious illness or the death of a breadwinner was experienced. As a charitable organisation, the society was reticent in public about its work, which was helped by contributions from the company from time to time. It had over 900 members in 1928, and had operated for many years. Another charitable activity which operated throughout the currency of the *Radiator* and no doubt beyond was the

6.3 Motor cycle and car club at Rest-and-be-Thankful (from Radiator).

Cot Fund, which raised money for the Sick Children's hospital at Yorkhill and other institutions by collecting tinfoil for recycling and through various small scale retail ventures. There was no shortage of good will.

One of the more surprising organisations revealed by the *Radiator* was the Night Shift Choir, which practised at 2.30 in the morning. It was still active 20 years later, and some of the medals it won in choir competitions still survive today as proof of its quality and dedication.

Aside from being a mirror of the social and charitable life of Albion, the *Radiator* was an outlet for the creativity and humour of the employees. One of these regularly produced poems in broad Scots, under the pen name 'Wigan Bill'. His first effort, given below, describes the experience of an Albion driver while collecting a completed chassis from a bodybuilder's premises south of the border:

A Driver's Lament

Of a' the toons tae which Ah've gaun–
Ah canna name them a' aff haun–
There's wan Ah railly canna staun,
That's Wigan.

Whene'er tae lift a bus Ah gang,
The builders they don't care a haung,
They simply say 'we'll no be lang',
At Wigan.

When Ah appear at five tae eight,
The foreman says 'Ye'll have tae wait,
There's twa three things we maun pit straight'
At Wigan.

An then he'll say 'as sure's Ah'm leevin,
Ye'll get this bus aboot eleeven',
But oftentimes it's nearer seeven,
At Wigan.

Bit never mind, why should Ah fret,
Awa Ah ultimately get;
When on the road, Ah shin forget,
Aboot Wigan.

Among Wigan Bill's other efforts were 'Do You Know It?', a thinly veiled and affectionate sketch of some of the characters in the Finishing Off Department, or 'Auld Tam'.

Many of the poems, jokes and anecdotes concerned either the workplace or Albion products, which were both held in high esteem. The anonymous 'Reflections of an Old Timer' is a case in point. Its opening verses are given below:

Reflections of an Old Timer

Way back some eight and twenty years,
With sheer grit and ambition filled,
Two clever gallant engineers,
Resolved a motor car to build.

Their task was hard you must admit,
The industry had just begun,
I'm sure with joy their faces lit,
When they saw built the first A1.

Hearts filled with pride each took his seat,
But joy was quickly turned to pain,
Their 'masterpiece' stopped in Finnieston Street,
And they had to push it home again.

Filled with resolve to win, not fail,
Both determined to see it through,
They very quickly had for sale,
A very much improved A2.

The poem went on to extol the Albion A3, A6, A10 and subsequent vehicles, concluding with a pat on the back for Murray and Fulton.

Team Work: A Story Without Words

With acknowledgments to an unknown Artist.

6.4 'Team Work' (Radiator).

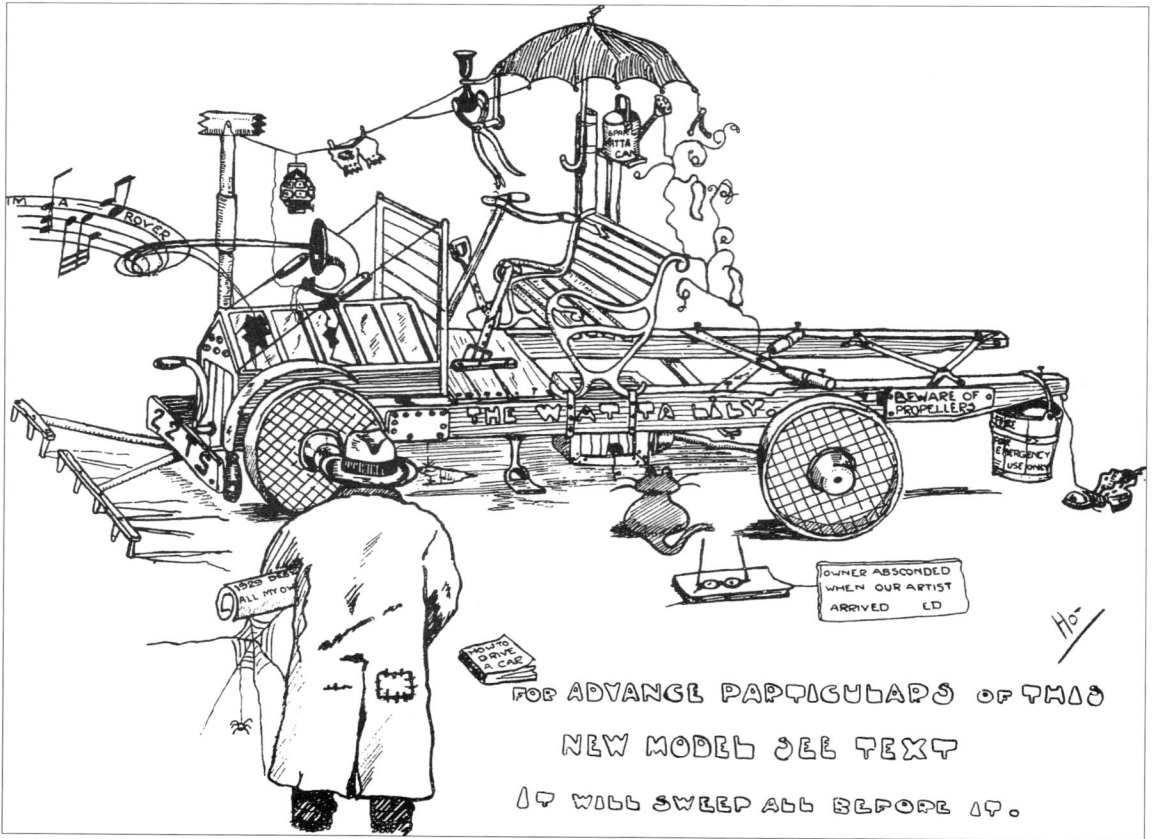

6.5 'The Watta-Lily', 1929 (from Radiator).

The fame this firm has gained and kept,
Was not attained by sudden flight,
The 'Heads', while their companions slept,
Were toiling upwards thro' the night.

The *Radiator* was an important means of routine communication both between the management and employees and between employees themselves, and also served as a general notice board. Occasionally serious business would briefly slip though, when, for example the managing director passed on a short New Year message which might contain a reference to the economic climate, but instances such as this were rare. The one other notable instance of business communication concerned the Suggestion Scheme, which remarkably by today's standards was still in place several decades after its introduction and remained so for several more. Details of numbers of suggestions and awards was regularly provided; the scheme was clearly popular. On a sadder note, the magazine was also used for obituaries of ex-employees. Bonus awards to apprentices, notices of retirements and the like were also regular features, as were pieces of news from the depots.

The *Radiator* also had a didactic purpose. Pieces on 'Monel Metal', 'The First Tyre', 'Higher Education in Great Britain and the United States' or on 'The High Speed Compression Engine' displayed a desire to improve the engineering knowledge of employees. An attempt was also made to advise employees of some of the management techniques in use in the business, and articles on 'Rate Fixing', 'Market Research' and 'Budgeting and Business Forecasting' were supplied for the general reader, revealing that the company was attempting to be as up-to-date as possible in the office.

In among these articles were to be found lighter pieces put in for sheer entertainment, such as 'Meandering on Mull', 'Travelling Companions' (a short story by Ian Chalmers), 'The Other End of the Telescope', by J. D. P., or articles, three in total, on 'Camera Work'. Halfway between business and pleasure were articles on countries where Albion had found markets, such as China, Canada or Africa, where vehicles were in use in unusual and exotic circumstances. Occasionally, too, a short piece on a new model, together with a photograph, would be provided, both to inform and to instil pride. All of this material combined with reports on lectures at the Institution of Automobile Engineers, cartoons, letters from former employees, reports on nights out and the humorous 'Radiations at Random'. This was the tailpiece to the magazine, contributed by the editor, and consisted of a series of jokes and puns of variable quality, some alluding to developments in the clubs.

The *Radiator's* heady mix bespoke a company that, at the time at least, was undergirded by high morale in its employees, partly resulting from the provision of good facilities and good conditions of work. These are not on their own sufficient to explain the pride and good humour that still reverberate from its pages. Readers would be foolish not to conclude that this had something to do with good management, in particular that emanating from the kindly, humane disposition of N. O. Fulton. It also had much to do with the natural cheerfulness of the working class Clydesiders who made up the majority of the workforce. These had been 'Albionised', to use a word found in the *Radiator*: imbued with the spirit of an organisation that had already become an institution, and in love with the products they made with their own hands.

CHAPTER SEVEN
The Turbulent Thirties

The 1930s was a period ushered in by an unusually long and savage economic depression of worldwide proportions and ushered out by the outbreak of the Second World War. Both of these cataclysmic events would inevitably affect Albion very deeply, but there were other important factors that turned the years in between into a period of great turbulence. First, there was dynamic change in the world of transport, and second, a gradual shift away from the certainties of management under the firm's original directors.

As the year 1930 progressed, the depression began to descend. Demand fell across industry as a whole as the worldwide slump signalled by the Wall Street crash of October 1929 began to bite. For the rest of the world, the depression could be explained readily by the working of the trade cycle and the withdrawal of dollars from trade. The rapid increase of output, prices and stock values of the late 1920s were there to be interpreted as precursors of the collapse. For Britain, there were other predisposing factors: an overvalued currency linked to over-large export industries. The slump was to last for three years, and while it proved bad enough in Britain, it turned out to be even worse in France and Germany. The demand for new transport was naturally affected, and Albion's sales of 1,302 vehicles for 1930 represented a reduction of about thirteen per cent on the previous year. In spite of this, there was no sudden ejection of surplus manpower, which remained at around 1,200 in the works and 600 on the staff, notwithstanding a note in the board minutes that at the end of 1930, the company was losing £1,600 per week. Overall, however, a post-tax profit of £50,286 was made for the year.

The company at the opening of the new decade was determined to continue the process of rationalising production that had started on the retirement of David Keachie the previous year. An attempt was made by William Pate to introduce a form of line production in the fitting and assembly departments. The finished part stores was reorganised, and in several of the machine shops a system for inspecting parts as they were being produced was begun, which reduced scrap. John MacLeod, an ex-Albion employee, was brought back from Austin, where he had worked as an efficiency engineer, and several similar appointments were made. Equally radical changes were instituted in the drawing office. Thomas Webster, the chief draughtsman for 21 years, in September 1930 had his contract of employment terminated at the age of 59, ostensibly taking 'retirement' because of overstaffing. Webster had worked for Mo-Car with Murray and Fulton in the

7.1 An Albion on the Glasgow/Walkerburn route, c 1930.

late 1890s, but was clearly part of the drive by the Pates to inject a new approach. This must, however, have had the support of the older directors. Webster was succeeded as chief draughtsman by Malcolm Beaton, who was promoted internally. He was to remain until 1916.

If turnover had suffered in 1930, it was by no means a disaster, and during the year there had been substantial orders from the War Office, the India Office and the Post Office, and the honour of an order from the King for a 30 cwt. lorry for the Balmoral estate. Building for the future took place through the opening up of an office in Calcutta, and J. D. Parkes spent six months in India on a sales trip. H. E. Fulton likewise spent four months in South Africa, and an office was opened in Johannesburg. In terms of new models, a new 2 ton chassis was brought in, and the Albion 'Victor' light bus was introduced for country services. The 'Valkyrie' 32 passenger bus, designed for maximum operating economy, was exhibited with the Victor at the Scottish Motor Show.

A further step backwards was taken by yet another member of the old board: J. F. Henderson, who had been a joint managing director with the two Fulton brothers, stood down but retained a seat on the board. G. M. Young, director and London manager became a joint managing director at this stage, as did George Pate.

The year 1931 has much in common with 1930. Business carried on in a depressed state, but the vehicle output of 1,271 units was very little below the 1930 total. Savage price discounting to maintain sales brought about a loss for the year of £18,966, although the bank balance was in credit to the tune of £29,717. As previously, the results hid some significant contract successes with

7.2 Albion trailer set, c 1930.

bus companies, the War Office, and railway and petrol companies. Several trucks were launched, uprated from previous models and with higher carrying capacities, and the 'Valiant' bus, capable of seating 32/36 passengers, was launched with a six-cylinder engine.

Factory reorganisation continued. N. O. Fulton, in the *Radiator* in October 1931 affirmed that

> The Management is, therefore, determined to leave no stone unturned to cope with the situation. Designs, materials, also Shop methods and practices which have been considered good in the past must be reviewed in the light of today's requirements and where necessary modified, and we rely upon the hearty cooperation of all in our employment to assist us in our efforts. Our interests are common and everyone can help. Notwithstanding the difficulties, sound foundations are being laid, the full value of which will be apparent when business becomes normal again. Meantime let our motto be: 'A stout heart to a stey brae.'

In practice this meant that the assembly of chassis, frames and engines had been reorganised to reduce physical effort, production of heavier parts was moved to the ground floor of the concrete block, and plans were afoot to transfer chassis assembly the following year nearer the bulkier sub-assemblies. The fragments of evidence that survive suggest that management had tolerated some restrictive labour practices in the factory for some time, but that they felt this had to change. A half-hearted attempt to organise a strike the following year would come to nothing.

At directoral level, the reduction in the influence of the old managing directors continued when H. E. Fulton resigned on grounds of illness, retaining his seat on the board. In May 1931 the company re-named itself Albion Motors Ltd., dropping the word 'car' from its title at long last.

1932 was again a depressed year, but a higher output of 1,362 vehicles was achieved, with a lower loss of £11,902 made, and substantial orders were taken

7.3 John Francis Henderson (from a 1924 portrait).

from the LMS, LNER and Irish Great Northern railway companies. The Post Office was again important. For the future, an Edinburgh sales and service depot was opened, and an office and spares store was opened in Belfast. A Nottingham office was also started up. Four truck models were launched, and the 'Venturer' double deck bus, Albion's first, was exhibited at the Scottish Show. G. M. Young, who had only recently been made a joint managing director, was compelled by illness to stand down, retaining his seat on the board. As a harbinger of improving times for the company, it cleared the last tranche of its £200,000 debenture issue, which had been taken out in 1918.

An important technological change had also made its mark at Albion in 1932. The high taxation content of petrol had caused an interest to be taken throughout the industry in the use of diesel engines. These gave twice the miles per gallon that petrol gave, which was very important to long distance operators, whose vehicles could often travel 80,000 miles per annum. It was later revealed that George Pate had taken the view that the Exchequer would not tolerate the revenue loss that a wholesale transfer to diesel would cause, and that such a switch by Albion would ultimately prove useless and a waste of resources. Diesels were not, therefore, developed, but business began to be lost to competitors. By 1932 Pate, Canute-

7.4 Albion rigid six wheeler of R.A.S.C. Motor Transport Department at Hounslow Heath.

like, failed to reverse the tide of progress, and as a result, Albion was forced to offer proprietary diesels on most of its trucks and buses by 1932. Government did several years later increase diesel tax, but it did not affect the engine's popularity.

During 1933 there would be far-reaching changes throughout the transport industry that would make a deep impact on Albion until the end of the decade. Between 1928 and 1930, a Royal Commission on Transport had been investigating a number of issues affecting this rapidly changing and dynamic industry, which included an examination of the relationship between the highly organised and regulated rail transport sector and road transport. The latter was still in a state of near anarchy, and was operated by many independent firms and individuals competing bitterly with each other and with other modes of transport. In order to bring clarity and organisation to this state of affairs, the Royal Commission recommended that road hauliers should be licensed. No immediate steps were taken to bring this about. In 1932, thanks to the conservative government, a conference was called under the chairmanship of Sir Arthur Salter to further the matter. There were three principal areas of concern. The first related to the sharing of highway costs between the different classes of mechanically driven vehicles and the means by which this could be achieved. The second related to the regulatory framework appropriate to road and rail transport, and the third concerned the devising of an equitable way of allowing the two sides of the transport industry to compete whilst safeguarding the interests of trade and industry.

The Salter conference recommended new and higher duties for commercial vehicles, in order to redistribute the costs of roads in accordance with the wear and tear caused by different categories of users. Salter also proposed that a

7.5 Albion charabanc at Lincoln's Inn Fields, London, early 1930s.

licensing system for goods operators should be devised to cut out unbridled and uneconomic competition. Where there were excess transport facilities in any area and a licence was requested, it was recommended that the position be carefully reviewed. Effectively, this could lead to the rejection of applications, or the permission being given for a smaller number of vehicles than requested.

As far as the redistributive element of these recommendations was concerned, the Finance Act of 1933 brought in the requirement that vehicles with an unladen weight of less than two and a half tons should pay a cheaper licence fee and be allowed to travel at 30 m.p.h., whereas those vehicles that were heavier should pay a higher licence fee and travel at 20 m.p.h. In terms of the licensing arrangements flowing from Salter, the Road and Rail Traffic Act of 1933 provided for 'A' and 'B' licences (for carriers wholly or partly for reward) which were subjected to the test of 'need' before issuance, whereas the 'C' licence (for firms carrying their own goods) was freely issued.

These new measures ushered in a wholesale programme of model development across the industry designed to achieve the maximum possible carrying capacity in chassis both above and below two and a half tons. The incentive was even greater to users who could operate chassis under two and a half tons, because of the higher legal speed limit. This called for careful design and the use of special steels and light alloys, which enabled lighter chassis weight to be combined with greater strength. Indeed, the 1933 legislation explains a great deal of the model proliferation seen at Albion during the 1930s, which is inextricably linked with

7.6 33 seater Albion in livery of Brighton Safety Coaches c 1934.

the firm's fortunes during these years. The undernoted table gives an overview of what happened in this sphere over the decade.

Albion Models Introduced and Withdrawn

Year	No. Introduced	No. withdrawn	Year Increase/ Decrease	Cum. Increase/ Decrease
1930	2	–	+2	+2
1931	8	2	+6	+8
1932	4	3	+1	+9
1933	11	10	+1	+10
1934	6	5	+1	+11
1935	18	15	+3	+14
1936	6	3	+3	+17
1937	8	7	+1	+18
1938	7	9	-2	+16
1939	2	6	-4	+12
	72	60		

7.7 Model 69 Valkyrie in Strachan's of Deeside livery, early 1930s.

Half of the models introduced and withdrawn were phased in or out during the years 1933–1935, and this quite clearly had much to do with the legislation of 1933. The table also reveals that, by 1939, Albion had introduced 12 more models than were withdrawn. A perusal of the vehicles which came in over the period makes it clear that the 1933 legislation encouraged the firm to extend its offerings further up the weight range, since the new materials being used meant that trucks weighing over two and a half tons could now carry proportionately heavier weights, which must have been attractive to customers. From 1933 onwards, heavier trucks of four, six and eight wheel configuration, capable of carrying up to 15 tons, began to take their place in the range.

Model complexity had begun to multiply from the late 1920s onwards with the evolution of the 'overtype' vehicle. It was not suited to use in countries such as Australia, South Africa or India to which Albion still continued to send many units. With the cab being on top of the engine, it made conditions for the driver inside intolerable in hot climates, and bonneted models of most types of chassis had to be produced in parallel with overtype vehicles. The development of six wheeler vehicles, too, had affected both buses and commercials, and there were models of each in this format. The legislative changes of 1933 must have finally

7.8 Valiant Model M70 32 seater bus supplied to Red and White Services, Lydney, c 1934.

put paid to any residual notion that the model range could be restricted to two basic vehicles. A market, driven by the incestuous factors of legislation and customer preference, was calling the shots in a more demanding way than ever before, and Albion would have been foolish not to respond.

As a general rule, most of the models introduced in the few years up to and including 1933 were withdrawn within a year or two of introduction, and it is clear this came about because they did not meet the weight and payload criteria which had been introduced by the 1933 Act. It was perhaps predictable that the vehicle ranges designed in the light of the new legislation would last much longer. A salient example was the M127, a truck with an unladen weight which brought it into the lower taxation category and yet it could carry up to 6 tons. This was produced from 1935 to 1941, and was highly regarded in the market place.

None of this could have helped the company in its desire to rationalise production, and on many of the vehicles, three, four or five types of engine, gearbox and axle were being offered. Life in the design and drawing office areas at Albion was consequently hectic, with four designers, Stanley Shaylor, Archie Dunsmuir, Tom Bibby and Cecil McIntosh responsible for engines, gearboxes, rear axles and chassis, and front axles and chassis respectively. In 1933 there were special tensions, and James Watt, the chief designer since 1920, was sacked by George Pate after disagreements on designing lightweight parts for three-ton trucks. He was replaced by James A. Kemp, who remained until 1959.

7.9 Valkyrie 32 seater in livery of Dodds of Troon, September 1934.

Little has been said so far concerning Albion's bus business at around this time. Buses, too, had been the subject of the Royal Commission on Transport of 1928–1930. In its second report of 1929, it had produced recommendations designed to put some order into the free-for-all that had existed in bus transport for many years. A fairer system of competition was required, and the public deserved convenient, regular and safe services at reasonable fares. The Road Traffic Act of 1930 divided the country into traffic areas, eleven for England and two for Scotland. In each a permanent, impartial commissioner was installed, assisted by two part-time colleagues chosen from names submitted by local authorities. The commissioners licensed drivers, conductors and vehicles. Drivers had to pass a special driving test, and their hours of duty were subject to an upper limit. Stage, express and contract journeys were distinguished, and licences to operate were strictly controlled by commissioners, thus providing a barrier to entry into the industry. Public need for a new service had to be demonstrated. The effects of this legislation may be judged by the numbers of buses and coaches on the roads of Britain during the 1930s. In 1930 the total stood at 52,648, and was at 53,005 by 1938. The legislation appears to have stabilised the industry, and also stabilised the demand for buses, although numbers put in and out of use fluctuated from year to year.

The absence of sales details over the 1930s permits only a glimpse of the mechanics of Albion's bus business, but it is safe to conclude that it was a mainstream player in the market. Its vehicles were represented in two of the largest Scottish bus fleets, Scottish Motor Traction and Western SMT. In 1933 SMT had 550 vehicles and Western 690. Albions were also represented in Glasgow Corporation's fleet, and in a number of fleets south of the border.

7.10 A 6 W Valkyrie in Young's livery at work in Paisley, 1933.

A good picture of the operation of the industry as experienced by a major user of Albion vehicles was provided in *Commercial Motor* of 3 August 1934. It featured Young's Bus Services of Paisley, which was the largest independent bus operator in Scotland. The journal complimented Young's for having built up a strong business in the Renfrewshire and Paisley areas, two locations that had suffered from the depression in the shipbuilding and associated industries. It picked out two factors as being of special importance, Young's system of maintenance 'that leaves practically nothing to chance' and 'discrimination in the choice of buses employed'. Young's had 82 Albion and two Dennis buses, the latter being 20 seaters. 13 of the 82 were double-deckers, and 11 of those had diesel engines. Their seating capacities ranged from 48 to 60 seats. The single deckers were principally 32 seaters, and eight of these had diesel engines. The annual distance covered by the fleet averaged about five million miles, or about 60 thousand miles per bus.

Young's buses, in their familiar red and yellow livery, plied routes from Glasgow to Largs and West Kilbride via Paisley, Johnstone, Milliken Park, Howwood, Lochwinnoch and Dalry. Their competitors were SMT, Western and Glasgow Corporation at different points on their routes. A weekly unlimited mileage ticket between Glasgow and Johnstone could be purchased for four shillings. Young's

7.11 A Viking Six in Young's livery leaving Clyde St., Glasgow for Largs, October 1931.

also undertook excursions, tours and private hire work, and kept a number of vehicles in reserve for rush periods and weekend traffic.

The firm's main garage and maintenance depot at Johnstone covered two and a half acres and had eleven vehicle pits. Vehicles were inspected and given a minor overhaul every four or five weeks (7,000 miles), and a complete overhaul every 70,000 miles. If there were symptoms of malaise with the vehicle, an unscheduled visit to the garage would naturally be arranged. Drivers were, where possible, assigned the same vehicle to enable them to become accustomed to its running and acquire a feel for its mechanical health. Length of service, however, entitled some drivers to machines with six-cylinder engines or servo brakes, which were very much preferred.

Commercial Motor admired Young's diagonally offset maintenance pits, which allowed mechanics to change adjacent wheels on different vehicles without getting in each other's way. The 7,000 mile service took about one day, and involved a mechanic, electrician, coachbuilder and two apprentices. A form listing the various maintenance operations was ticked off as they were completed. In addition to the 7,000 mile service, diesel engined buses were given a weekly oil and fuel filter service. The buses were all greased, washed and cleaned internally on a daily basis. The main overhaul area had specially configured pits with end bays shaped for work on engines, gearboxes and back axles, and Young's were equipped with the best electric test equipment and charging plant and also a Bosch spark plug tester and an eight-ton bus lift.

Commercial Motor found that one Albion Valiant machine had covered 175,000 miles in two years at an average petrol consumption of 7.3 m.p.g. and an oil consumption of 1,360 m.p.g. Three Albion 55 seaters with Gardner diesel engines

7.12 Albion bus on tilt test, c 1935.

which had recently been acquired, had covered 9,000 miles in about five weeks and had given 12.5 miles per gallon. The brakes had not been refaced, the injectors had not been changed and there had 'not been one involuntary stop'. Young's possessed a large range of Albions which included the Valorous six-wheel double decker, the Viking Six single decker, the new Venturer double decker, and the Valkyrie six-wheeler, 39-passenger single decker. In addition to the fleet of Albion buses they also owned a number of Albion trucks, operated through a haulage subsidiary.

The momentous disruptions that had taken place in the economy and those that subsequently flowed from the changing regulatory framework in which road transportation operated called for an unprecedented intensity of management attention, perhaps only paralleled during the slumps and disruptions of the early 1920s. And yet, at this time, the departure or partial withdrawal of the older directors continued unabated. N. O. Fulton in June 1933 resigned his position as joint managing director, but retained his seat as chairman of the board of directors. Having presided over the completion of the substantial works reorganisation, which he had fully supported, he clearly felt that the time had come to step back. At the same time H. E. Fulton, as a result of continued ill health, resigned from the board. In 1934, Walter McFarlane, Company Secretary, was elevated to the board, but Robert Thomson resigned on 30 September. A momentous and stunning change occurred on 27 July 1935, when N. O. Fulton

7.13 N. O. Fulton.

died suddenly at his home, St. Edmund's, Milngavie. This event had the effect of consolidating the position of George Pate, who was appointed chairman.

In losing N. O. Fulton, the company had lost a Solomon. An 'Appreciation' in the *Radiator* by McFarlane makes the genuine sense of grief felt across the whole company quite apparent. He spoke of Fulton's 'wise and considerate leadership' and 'the great regard in which he was held throughout the organisation'. All admired 'his gifts as an engineer, as an organiser, his energy, his rectitude, his loyalty, his steadfastness, and amidst all the cares of his high office his forbearance and consideration'. McFarlane noted that

it was given to the great ones of earth by the magnitude and success of their operations to command our respect, to seduce our admiration, but not always was it given to them to win our love. Where such is the case, success must surely be something of a mockery. Without the friendship and goodwill of those around us, especially of those who look to us for support and guidance, life loses the greater part of its charm. When the Creator formed the heart of man he planted goodness there as a principal element of character, and as a means of attracting the affection of others, and when high responsibility supervenes on this, so far from detracting from it, it ought to enable it to diffuse itself more extensively. Thus did it act in the case of Mr. Fulton…Possessed of infinite tact, he would investigate the numerous little human problems that came before him

113

7.14 Albion bus in Midland livery, mid 1930s.

with the utmost patience, assisting here, advising there, with a solicitude and care scarcely to be expected from one so fully occupied in other directions.

J. F. Henderson joined in the tribute, noting that Fulton had 'earned the admiration and respect of all', and that the success of the company had to a large extent relied on 'his great organising ability, sound judgement and energy'. It cannot be doubted that morale and industrial relations throughout the works and office, if never perfect, owed their health and stability to this wise and kind man, whose attributes are more than confirmed by his portrait. The proof was visible at his funeral at New Kilpatrick Cemetery, where a multitude of friends and 'a large representation from the works and office' took the trouble to pay their last respects. McFarlane's final word could not have been more apt: *Si monumentum requiris, circumspice.*

At the same time as George Pate became chairman, H. W. Fulton, BSc was appointed to the board. N. O. Fulton's elder son, he had gained a first in engineering at Glasgow University, receiving his practical training with Harland and Wolff, Scotstoun, John Brown & Company, Clydebank, and then at the Bergius Company of Glasgow, the latter a firm of engine manufacturers. In 1928 he had undertaken a world tour, which included works visits in the United States to many of Albion's machine tool manufacturers as well as to motor companies in Akron and Detroit. He had visited Albion agents in New Zealand and Australia. On returning, he had entered Albion to gain experience in various departments and in 1931 had become Assistant Technical Manager, in charge of the Technical and Experimental Departments. He had become Technical Manager early in 1935, keeping responsibility for this function as a director.

At this point the board consisted of George Pate, chairman and managing director, John Francis Henderson, deputy chairman, G. M. Young, J. D. Parkes, Alex Donaldson, H. W. Fulton, William Pate and Walter McFarlane, who was also

7.15 Albion stand at the Rand Motor Show, April 1936.

Company Secretary. The Albion board minutes disclose a subtle change in the operation of the board over the 1930s.

At the beginning of the period, a typical monthly board meeting might begin with the reading of the last minutes, business arising and apologies, then deal with 'exceptional items' of a non-routine nature, if any. An example of this might be a resignation or a major non-recurring problem. Share transfers would then be authorised, and a series of routine monthly reports would be produced and discussed. First would come sales invoiced in the month, duly analysed. Next, orders received for the month, on which comments would be passed and comparisons made. After this, business prospects would be presented, and then works output was discussed, followed by a report on 'the development of new types', which would discuss the performance of prototype vehicles on trial. A financial statement followed (initially the responsibility of H. E. Fulton). This would consist of bank balances, the outstanding debenture balance and stock levels, expressed in units and where applicable, the forthcoming dividend. A half-yearly balance sheet was also examined. From the middle of the decade, reports became a little more detailed, and a 'technical manager's report' succeeded the report on 'the development of new types'. This also covered the level and cost of warranty claims, and was presented by H. W. Fulton. Vouchers were now provided to support bank balances (presumably in the form of statements or

7.16 Albion Model 34 at work in South Africa, late 1930s.

reconciliations), and short-term cash flow forecasts, none of which have survived, were presented. An impression of closer scrutiny, perhaps a reflection of Pate's intense and determined personality, emerges.

Prior to Pate's chairmanship in 1933, a loss of £6,463 was recorded on sales of 1,380 vehicles, reflecting the last year of recessionary conditions. In 1934, profits of £44,681 were made on sales of 2,003 vehicles, sold to old customers such as Harrods, Bovril, Lyons and the railway companies, as well as to Carter Paterson, who purchased 75 machines, and the Post Office, who took 64. That year, the company felt the benefit of pent up demand, which had resulted from the uncertainty surrounding Salter. Confidence in the rising home market caused the company to hire additional depot salesmen, taking the total up to 60 from 44. 1935 was a celebration year. A record 2,918 sales units was achieved, and a profit of £82,057 made. A contract was won from the Air Ministry and a further prestigious truck order from the King was received. In 1936, further plant additions were made and the night shift was enlarged, to keep up with production, which rose to an all-time record of 4,051 vehicles. Profits were £148,576. A bonus issue of shares was made of £82,900.

Albion was again booming, and new heat treatment, gear cutting and laboratory facilities were started in 1937, when a service depot was opened in Johannesburg. Extensions were made to Sheffield depot, and a new depot opened at Leeds. Profits on 3,968 units of chassis sales during 1937 were £186,049. Albion's own design diesel engine was now launched. A minor blot on the business landscape was a national apprentice's strike, which affected many engineering employers on Clydeside. It was duly resolved, but had resulted in lost output. In 1938, 3,752 vehicles were sold at a profit of £198,689. A sales office was opened at Hull. Albion's post-Salter vehicles had proved a huge success.

7.17 Chassis department, Scotstoun, 1934.

7.18 Machine department, top flat, Scotstoun, 1934.

7.19 Bob Coutts delivering a chassis to bodybuilders in Portsmouth, 1937.

As a tailpiece to the decade, the uncertainties in the international situation caused by Hitler had induced caution in buyers for the first time in five years, and output fell to 3,102 units, which generated a profit of £174,561 for 1939. The balance sheet showed assets of £1,707,635 that year, as against the figure of £978,265 in 1933.

Apart from the very positive achievements of the 1930s in works expansion and reorganisation, record output levels and the profits associated with Albion's completely regenerated and expanded model range, there were many other small but significant developments that are worthy of mention. The first of these was the addition of the motto 'Sure as the Sunrise' to Albion's publicity material. This would invariably be superimposed on the rising sun emblem that for a number of years had adorned the radiators of Albion vehicles, an early example being a 1934 advertisement for the Venturer, Valkyrie and Victor buses. The secret which lay behind its successful long term use was its apparent simplicity. By invoking the most prominent example of daily reliability in nature in a verbal way, a distinctive visual device was transformed into a powerful symbol of dependability. The word 'sure', which consisted of letters common to the word 'sunrise', slipped easily over the tongue and helped make the phrase memorable. It became synonymous with Albion and struck the nation's imagination.

While Albion radiators were assuming a greater significance in advertising terms, the magazine that was named after them died in 1937. The last edition

7.20 H. E. Fulton.

contains no explanation for its demise, but in the absence of N. O. Fulton, its mainstay of support had gone. Its disappearance seemed to symbolise the accession of the new leadership of the board, who must have been keen to make their own impact on the firm.

On 8 February 1936, links with the old board had been weakened when H. E. Fulton died at his home in Bearsden. An obituary in the Albion log book ascribes the expansion of the sales organisation both before and long after the Great War to his foresight and wise guidance, and it was he who had masterminded and managed the repurchase of War Department vehicles for repair and resale in 1919/1920. H. E. Fulton, too, had been blessed by the charm, courtesy, integrity and sincerity that had marked out N. O. Fulton as a prince among men, and had also possessed the sense of care for all Albion people that characterised his brother.

In 1935, with a view to possible further expansion, Albion had purchased the Yoker premises of the Halley Company, which had ceased existence that year. Halley vehicles had acquitted themselves well in the first world war, and like Albion, had tried to adopt a 'one model' policy in the post-war period, designing

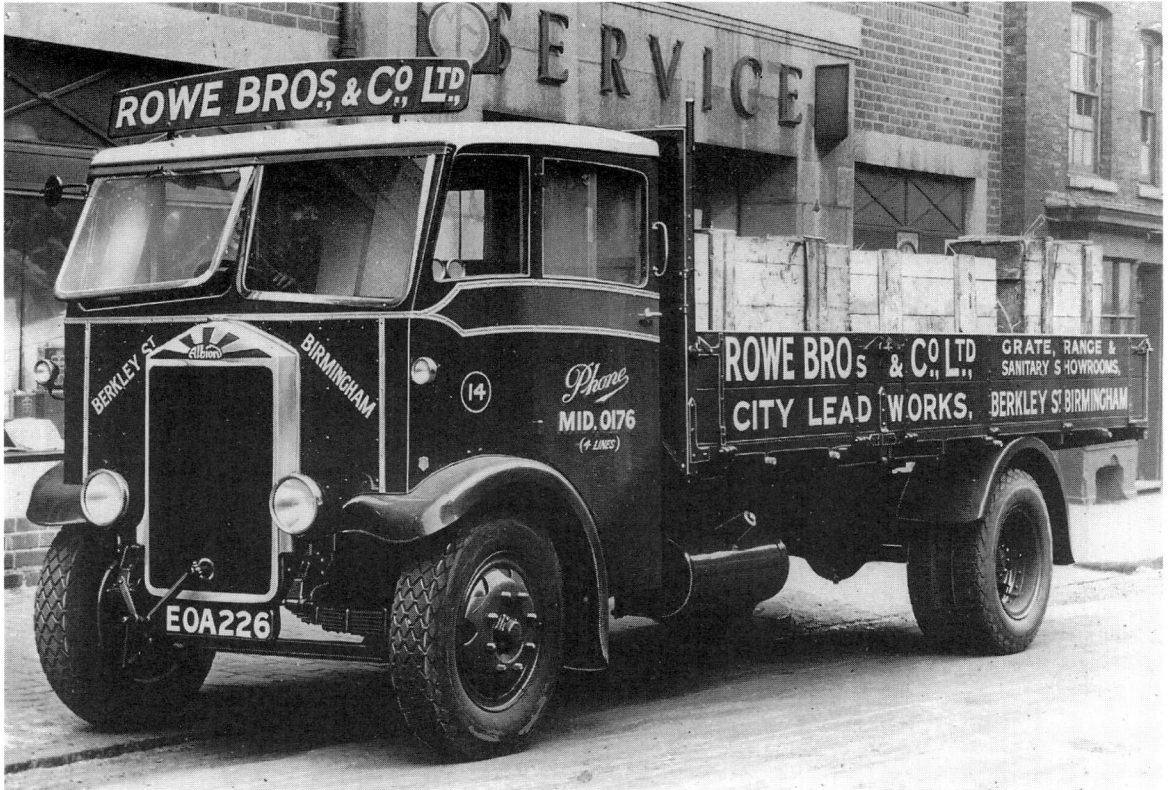

7.21 M127 5$\frac{1}{2}$ ton lorry, built 1935-41.

a chassis that would take three and a half ton loads or bus bodies to seat up to 29 people. This was slowly developed, but was not sufficient to sustain the factory when ready, and so older models were re-introduced. George Halley had died in 1921, and unsuccessful trading thereafter brought the company into voluntary liquidation in 1926. North British Locomotive company took a fifty per cent stake at this stage, but by 1934, decided to end the losses that continued to be made. Albion's main interest was in Halley's property, which, as was noted, consisted of a relatively new purpose-built factory. By 1935, therefore, Albion had ended up being the only survivor of the first phase of the early Scottish motor industry, and in possession of two out of the five purpose-built factories that it had spawned.

Before the start of the war, which had been widely anticipated, Albion had been making preparations. Additional ambulance and fire protection provision had been made at the cost of £1,000, and a board decision had been taken to make arrangements for air raids completely independent of neighbouring premises in order to protect Albion employees best. A budget of £15/30,000 was established for the erection of three groups of air raid shelters close to the factory, capable of taking all Scotstoun employees, much of this recoverable. The top flat of the Yoker offices was taken over by the government, and Albion moved duplicate copies of

7.22 Model T561 14 ton Gardner engined van for Melrose's Tea, 1937.

7.23 Model T561 with Gardner engine, 1938.

all tracings to Yoker for safe custody. The Lincoln depot was commandeered by the government, and the sales office there relocated. Staff in the London office were transferred out to the Service Depot at Willesden.

Meantime, on the eve of the war, Albion had been losing manpower to neighbouring firms involved in armaments contracts. Albion's payment record was good. Most employees had for decades been on bonus schemes, for which N. O. Fulton had largely been responsible. Details have gone, but those in the works operated on a piecework system. Those not covered by this were given an annual bonus based on a percentage of pay related to company profits, payable at the annual holidays. This was available in three bands, and Albion had frequently paid out on a higher percentage band than profits justified. These measures were not in 1939 sufficient to prevent employees going to neighbouring firms. 25 per cent of toolroom manpower had resigned, and skilled turners and inspectors had left in numbers. George Pate ascribed this to the cost-plus nature of the government contracts that were being fulfilled by these firms, which, he claimed, enabled the firms affected to pay unduly high wages and poach good workers. Albion were at this stage still producing predominantly for commercial customers.

Some years earlier, the Luftwaffe had secretly acquired aerial photographs of Albion's Scotstoun plant, and unknown to the firm, a Damocles' sword hung above it.

CHAPTER EIGHT
World War Two and After

On 30 January 1933 Hindenburg appointed Adolf Hitler Chancellor of Germany and thereby dramatically increased his influence. The Austrian Chancellor, Dolfuss, was murdered by locally based Nazis on 25 July 1934, and by 2 August Hindenburg had died, whereupon Hitler became an absolute dictator. On 3 October 1935 war between Italy and Abyssinia began, and ineffective economic sanctions were announced against Italy on 18 November by the League of Nations. The Rhineland was remilitarised on 8 March, and on 5 May Italian troops occupied Addis Abbaba. On 13 March 1938, Germany annexed Austria. On 28 September, the British Navy was mobilised, and one day later the Munich agreement was made between British Prime Minister Neville Chamberlain and his counterparts, Daladier, Hitler and Mussolini.

But Hitler was not appeased: on 16 March 1939 he annexed Bohemia and Moravia, and on 28 March, an anti-Polish press campaign was begun in Germany. On 27 April, conscription was introduced in Great Britain, and the next day, Hitler denounced the Anglo-German naval agreement and the Polish non-aggression treaty. In May that same year, Britain concluded a defensive agreement with Turkey, while Italy and Germany signed a pact. An Anglo-Polish treaty was finalised in London. When Germany invaded Poland on 1 September, 1.2 million people were evacuated across England and Wales. The next day compulsory military service was instituted for all British men between 18 and 41. War was declared between Britain and Germany at 11 a.m. on 3 September 1939, operational as from 5 p.m.

This time, the British government was much better prepared for the control of war than the first world war had found it. From the early 1930s on, government departments had been deeply engaged in war plans, which were put into action within a few days of the war's commencement. Schemes were put in hand quickly for the reservation of occupations, to prevent the call-up of key workers, the Ministry of Labour stood ready to retain the goodwill of the work force, and there were schemes for food rationing. There were of course significant differences between this war and the last: much of it would be fought out on non-European battlefields, and the airborne attack would become a major feature, capable of visiting terror and mass destruction on the civilian population as well as the military.

The opening months of the war at Albion were characterised by two outcomes of the war planning process that had not gone as smoothly as they should have

8.1 BY3 truck for the Ministry of Supply, 1940–41.

done. Remarkably, there was no serious rationing of shipping space by this time, and the authorities responsible for ordering military vehicles had not made up their mind concerning the types they wanted Albion to supply. What was certain was that Albion was seen as a specialist producer which would have been wasted supplying run-of-the-mill three ton load carriers, and indeed these types of vehicles tended to be ordered from the firms in the south that over the 1930s, had learned to more or less mass produce undifferentiated vehicles of this kind. Before the government orders came in, therefore, it was still possible to send exports out, and these included a number of vehicles purchased by South African Railways. Even these were beginning to be dogged by bottlenecks at bodybuilders, shortages, and dramatic price increases in components. Within a few weeks of the declaration of war, Mr. Donaldson, purchasing director, was having to report that Dunlop wheels had gone up by four per cent, Hardy Spicer joints by two and a half per cent, valves by seven and a half per cent, Girling and Lockheed brakes by seven and a half and five per cent respectively, while S.K.F. bearings had increased by 15 per cent. Aluminium pistons were another problem area, at seven and a half per cent.

Albion's military vehicle experience was, of course, considerable. There were still staff around who had been involved in the legendary A10s, and a number of vehicles had been despatched to India for military purposes throughout the 1930s. By 1939 there were already British government contracts in hand. Between 1933 and 1937 Albion had supplied the army with quantities of the M131 vehicle, a three ton, six wheeler, and from 1935 to December 1940, the M126 vehicle was supplied as a military ambulance. When the outbreak of war came, one of the government's first actions was to commandeer all M127 vehicles in Albion's shops and in civilian hands. These were highly successful, light and reliable vehicles capable of carrying up to six tons and were in overtype format.

Once orders from the military started to come through at the turn of 1940, it became progressively more difficult to carry on with civilian business, although a trickle of civilian work permitted by the authorities meant that it was not completely eliminated. Early in the war the government made it clear that the only exports allowable were of spares, which helped resolve priorities and brought about the cancellation of orders. Quantities of vehicles provided to the government over the war years are no longer available, but the main range of models provided is given in the undernoted table.

Model Type	Dates Produced	Description	Order Sizes
BY1	1/38–5/40	3 ton overtype	N/A
FT11	2/40–8/44	3 ton overtype 4 x 4	500
CX6	6/40–6/41	10 ton bonnet type	N/A
BY3	6/40–11/41	3 ton overtype	500
CX23	3/41–4/44	10 ton overtype	N/A
BY5	5/41–8/45	3 ton overtype	758
CX24	8/41–8/44	20 ton tank transporter	350
CX22	11/43–6/45	gun tractor, bonnet type	N/A
CX3	7/37–?	7¼ ton overtype	N/A
CX4	1/38–?	7¼ ton bonnet type	N/A

It cannot be claimed that the list is comprehensive, indeed there were experiments afoot with various other types of vehicle at Albion throughout the war, notably with the eight-wheel drive vehicle which is illustrated, the CX33. Another experiment carried out at Albion was the powering of trucks by 'producer gas'. This involved fitting the vehicles out with large peculiar-looking 'gas bags' which ran the length of their roofs. This was tried on the CX vehicles, which were able to run on 65 per cent power using this substitute fuel. The BY5 vehicle, too, was altered as a crane-carrying vehicle. The range provided by Albion extended from the most prosaic to the most innovative and complex kinds of vehicles, even encompassing folding boat transporters.

In addition to the many types of vehicles that were built and tested at Albion, there were several special jobs in progress during the war years. The first of these was the production of revolvers and sten gun barrels at Yoker. This was achieved

8.2 Ambulance AM463 as used by the Air Ministry.

by means of a special type of contract, wherein Albion was paid a rental fee for the use of the Yoker factory, with the government providing reimbursement of all costs and a manufacturing fee of four shillings per revolver, plus an undisclosed amount for each sten gun barrel. In addition to this it was initially proposed that Albion should get an annual construction fee for the revolvers, to cover the management time, of £8,000 per annum. This was subsequently reduced to £2,500. At a later stage, when the government provided the necessary working capital for pistol production, the fee per unit was dropped by six pence. In addition, there was a contract for torpedo engines, which was carried out at Scotstoun. This was well under way by October 1941, at which point there were 100 people and 100 machines employed on this venture, and by May 1942, the first torpedo engine was being tested. The torpedo engine contract was placed on a similar basis to the revolver contract, which was referred to in the Albion board minutes as an 'agency' arrangement.

A feel for Albion's wartime working is still available from the surviving documentation. As ever, there was a shortage of labour, and women were again drafted in. Out of 1,700 employed at Scotstoun by May 1941, 50 were women, but by the next month this had gone up to 300, which may have a factor in the strike which lasted for over a week in the gear cutting section at around this time. By September 1941, there was an order book of 2,676 vehicles outstanding, about a year's work, and modifications to the working week had been put to the

8.3 BY5 lorry with folding boat equipment, 1941–45.

'workspeople', as the board minutes so quaintly described them. By the end of September, 420 out of 1,790 at Scotstoun were women, and by the year end the figures had risen to 600 and 1,960, with a further 633 women and about 100 men at Yoker. The women did all manner of work, in the toolroom, in assembly and of course on torpedo engines and the pistols at Yoker. A shortage of setters arose and took some time to resolve, and shortages of parts were felt early because of the growing demands of aircraft construction. For the five months up to July 1942, a three-shift system was in operation. Key parts that could not readily be sourced from Britain, particularly buffer brackets, spring seats and stub axles were being imported from the United States. There were eventually shortages of steel.

At Yoker, the Enfield .38 pistol manufacture was well in hand in 1941/42. Government supplied the specialist and other machinery needed, and Enfield supplied drawings, jigs and fixtures. The drawings, unfortunately, did not have dimensional tolerances on them, a practice, paradoxically, that had been set by the U.S. Armory at least 100 years earlier and which had been adopted by Albion from 1900 onwards. H. W. Fulton had been responsible for starting this operation up, and noted that this 'grievously hindered the task of producing these with the female unskilled labour that was all that was available'. It was impossible to know, except by experience, which degree of variation from the precise sizes on the

8.4 CX33 8 wheel drive tractor.

drawings was acceptable. In the event, fresh drawings had to be prepared by the Albion drawing office, to let work proceed, and help with assembly was given by 'a small band of elderly watchmakers whose accuracy and nimble fingers more than made up for their, in some cases, low physical condition.' As far as the sten gun barrels are concerned, the order had come about as a result of the late development of the British army's first sub-machine gun by Enfield. The main producers were the Royal Ordnance factory at Fazakerly and B.S.A. Co. Ltd, but many other producers and sub-contractors were involved in an attempt to expedite output, hence Albion's participation.

By December 1942, the orders for the pistols and sten barrels had been stopped, since adequate provision had by this stage been made. In January this had led to what the board minutes describe as 'action by the workspeople'. A deputation consisting of Major Jackson Millar, son of one of Albion's first shareholders, a Mr. Forbes and a Mr. Young of the Ministry of Production, Mr. Glenny of the Ministry of Labour and Major Lloyd, M.P. explained that the weapons were no longer required to the 'Shop Production Committee'. Unfortunately a state of affairs satisfactory to the military was a real hardship for the weapons workers, and when this was pointed out, it was decided that they should carry on producing meantime, until an attempt to find replacement work was made. By March 1943, the factory was taken over by Coventry Gauge and Tool for another military project not involving Albion and jobs were saved.

The Luftwaffe had of course done their best to prevent Clydeside munitions works such as Albion from maintaining production. On 14 November 1940 in the wake of the Battle of Britain, Coventry had endured devastating air attacks, followed by Bristol on 2 December. On 13 and 14 March 1941 it had been the turn of Glasgow, or more precisely Clydebank. Only one small and discrete building at Albion, the Experimental building, was hit, and while it had to be

128

8.5 Standard AM463 Air Ministry tanker.

demolished, the equipment inside was saved and there had been no injuries. A legacy of broken windows and roof lights on other parts of the factory remained for the rest of the war. The board minutes record soberly that 100 employees had their homes destroyed by enemy action, and that four staff and nine works employees had lost their lives. At Yoker, there had also been some damage. Once they had sorted out their affairs, the workforce who lived in Yoker and Clydebank had gradually drifted back to work. Albion's Glasgow workforce were not the first to suffer from the enemy's attacks. Incendiary bombs had fallen on Birmingham depot on 25 August 1940, although no damage of any substance was sustained. Willesden, though, suffered a fate more in keeping with London generally. On 15 September 1940, the entrance gates to the yard and neighbouring roadway were hit by a bomb and blown away, and damage from neighbouring houses damaged the depot's roof glazing. On 26 September, an unexploded parachute mine which landed close by caused its evacuation until 2 October, and on 20 October, an incendiary bomb landed on the roof without doing any major damage. On 22 and 23 December 1940, Manchester was raided heavily and incendiary bombs came through the roof of the depot, which was also damaged by blast and shrapnel.

As well as providing vehicles and weapons for the war effort, Albion was able to supply several senior managers who assisted at national level for long periods of time. On 21 July 1940, George Pate was contacted by no less than Max Aitken, Lord Beaverbrook, the Canadian newspaper tycoon and politician, who had been

8.6 Albion BY1s at Loughborough, 1939.

placed in charge of aircraft production by the wartime cabinet. Pate took over as Director General of Engine Production on 26 July, having discussed it with the Albion board beforehand. The board made arrangements for the extension of his service contract, since the period of his secondment was an indefinite one. Under the arrangements made, it was still possible for Pate to return northwards for monthly board meetings, and these he continued to chair. He left his new appointment on 10 April 1941. By this time the energetic (not to mention autocratic) Beaverbrook and his team had helped Britain create the aircraft needed to prevail in the Battle of Britain, which ended victoriously on 15 September 1940, having resulted in the loss of 1,733 German aircraft as against the loss of 915 sustained by the R.A.F. In view of the threat of invasion, all projects and resources that could be mobilised for this effort had been accorded absolute top priority, and Pate thus had his character tempered in the white heat of national emergency.

Later in the war, H. W. Fulton volunteered his services to the Ministry of Supply. Feeling that the production of war vehicles at Albion was well under way and being competently looked after by the senior staff at Scotstoun, he took the view that he should be playing a fuller part at a national level. He applied for and was given an Assistant Director's post in the procurement section of the ministry responsible for providing the military with vehicles of all kinds, reporting to Brigadier Hedges. He was granted leave of absence by Albion in July 1942, having seen the pistol and sten gun operation at Yoker successfully established. He found himself immediately thrust into a world with which he was very familiar: a number of motor industry executives were already at the ministry, and he found himself having to negotiate with and resolve disputes between supplier firms with which Albion had dealt for many years.

Among the more notable problems which Fulton faced was the co-ordination of common supplies between U.S. and British forces. In particular, American petrol, known as MT80, contained tetra-ethyl lead, which damaged British vehicle engines not designed for its use, and problems with burnt-out valves had been

8.7 AM463 articulated lorry supplied to the Air Ministry.

experienced in North Africa. Another problem concerned the different screw thread systems that were in operation in British and American machinery, and an attempt was made to resolve this by creating a common standard, the 'unified thread'. This was not a great success, and caused problems in the post-war period. At a more anecdotal level, Fulton was constantly frustrated at the inability of the ministry typists to handle the technological details in his reports, and eventually had his typist, Jessie Mackay, seconded to the ministry, where she appears to have been a great help in a whole range of ways.

While in London, Fulton was fortunate to have escaped with his life. On one occasion the lodgings which he shared with a Major Currie were destroyed by a VI rocket while he was absent, but Currie was killed. His memoirs reveal that this had happened to him before, but the loss of Currie and the randomness of fate both gave him a tremendous shock. From this point onwards, he was invited to live at Chalfont and commute to London by train.

Among the many routine tasks which faced the ministry staff was the development of an amphibious vehicle, loosely modelled on the U.S. DUKW, and also the search for substitutes for the rubber tyre, which were unsuccessful. More exciting was the opportunity to land in France behind the assault troops to glean what information was available on German technical development, as German laboratories and workshops were captured. In due course, Fulton traversed Belgium, Holland and Germany as the war drew to a close, and was one of the first to witness the full horror of Buchenwald. The board minutes record that he was occasionally able to attend board meetings, which no doubt he would combine with visits to his family on the outskirts of Glasgow.

The statistics of war, so far as Albion was concerned, were as ever impressive. 2,492 vehicles were completed in 1940, 1,860 in 1941, 1,213 in 1942, 1,456 in 1943,

8.8 FT11 4 WD truck for the Ministry of Supply.

1,038 in 1944, and 980 in 1945. The reducing numbers were a reflection of the relative complexity and size of the vehicles being tackled at Albion as the war progressed. In addition to these, 684 pistols and 31,550 sten barrels were produced in 1941, the figures for 1942 being 13,261 and 318,523. 7,477 revolvers and 80,644 barrels were produced in 1943, as well as 519 torpedo engines, 91 of which had also been produced in 1942. 621 torpedo engines were produced in 1944.

Profits at Albion during the war were, predictably, strong, and are given below together with sales:

	Sales	Profits
1940	£n.a.	£124,768
1941	£n.a.	£120,024
1942	£2,446,000	£125,564
1943	£3,027,000	£126,894
1944	£3,122,000	£121,654
1945	£2,588,000	£137,486

The profits are quoted after tax. Shortly after the start of the war, a government edict was issued instructing limited companies not to disclose tax payable, as a national security measure. Profits were much the same throughout the war as they had been in 1939, when tax of £65,000 was paid on profits of £175,992, and so the war can be seen to have preserved rather than greatly enhanced the company's earning power. War finances did, however, impose temporary strains as a predominantly cash-on-delivery business was converted to one paid by

8.9 WD CX 24 S tank transporter.

government on the strength of invoices, often involving investigations by government-appointed auditors. In May 1940, to ease the situation, Albion cashed £55,000 in corporation loan stocks, but there was in any case no shortage of working capital. The Clydesdale Bank had offered an overdraft facility of £200,000, at four per cent, early in the war, and this was used from time to time. By July 1941, the overdraft stood at £68,958. At the end of the war, however, there would have to be financial rebuilding. Depleted assets would require to be replaced, and the high levels of stocks no longer needed to meet the government's fluctuating plans would have to be brought down.

The war years, too, had witnessed other important changes, to some extent over-shadowed by the dramatic national and international scene. Not least among these was the death at the end of may 1941 of John Francis Henderson, the last of the quartet of original directors. From 1936 onwards, he had served as deputy chairman of the board, and for many years prior to that had been responsible for the bodywork and service operations at Albion. Henderson's portrait confirms the sharp intellect that had singled him out as a Thomson Scholar in Physics at Glasgow University in his youth and the dignified reserve that had fitted him for the committee work he had undertaken on behalf of several professional bodies associated with the motor trade. If not a pioneer in the sense that Murray and Fulton had been, together with H. E. Fulton he had been a stalwart throughout the company's first three decades and beyond. At around the same time, another long serving director, G. M. Young was forced to retire through ill health.

While the war had been taking its course, there had been highly significant developments at Albion that were having an effect on shop floor morale and causing concern to some of the directors. These related to the bonus scheme that had been in operation from before the great war. Although its particulars have gone, it operated on two bases, a piecework system for production workers, and a profit related system for those not covered. The latter part of the arrangements involved an annual percentage of bonus that was paid in bands, dependent on

8.10 BY1 petrol tanker for the Air Ministry.

profits achieved. Frequently a bonus band that was not strictly warranted had been awarded by the management. At a board meeting of 24 March 1942, George Pate announced that he was very dissatisfied with this part of the bonus scheme, and declared that he would be discussing the matter with the North West Engineering Trades Employers Federation with a view to changing it.

At the next board meeting it was decided in principle that it should be altered, and there was a general exchange of views as to how a replacement might operate. At the meeting of 12 May, 'certain members' of the board took the view 'that it was undesirable to make any change during the present hostilities', but in spite of this the board as a whole decided in principle to withdraw the existing scheme. At this stage William Pate and H.W. Fulton requested 'that their vote in opposition should be recorded'. The board next decided that the Albion Long Service Certificate, which came with a £25 award and increased the bonus entitlement, should be incorporated in any new scheme, which George Pate undertook to draft. In June 1942, the board agreed that only breaks for sickness or national service should be allowed under the provisions for operation of the long service certificate and bonus, which clearly had been the subject of more generous interpretation in the past.

8.11 6 wheel Albion WD truck.

In April 1943, matters had not moved much further forward, and the board minutes record that the bonus and related arrangements would soon be coming up for discussion. The next month, in Fulton's absence, it was noted with some alarm at the board meeting that the long service award's five year eligibility condition was excluding a large proportion of employees from participation, particularly women and apprentices. The matter was deferred. On 19 June 1943, again in Fulton's absence, it was decided without finalising any new scheme that the award for the year should be threepence in the pound for hourly paid workers and sixpence in the pound for staff. This was less generous than normal, and would only cost the company a few thousand pounds, whereas in 1936, some £15,000 had been involved. At a subsequent meeting of the board, it was reported that a 'certain amount of disappointment and some dissatisfaction' had been expressed by the workforce concerning the payout, but that 'it had passed off quite satisfactorily'. The shop stewards at Albion, who had been gaining strength and building up union membership since the arrival of John Gray as shop stewards' convenor about 1936, announced that they would be investigating the management's competence to alter the scheme. Under the wartime arrangements, the Ministry of Labour had put in place machinery to arbitrate on disputes and make awards, so as to avoid the difficulties that had arisen during the previous war. It should be emphasized that these changes to the bonus arrangements were effected when Fulton was not in regular attendance, and over a period when William Pate had been quite seriously ill. The other directors, clearly, were not prepared to stand up to George Pate on this question.

The board minutes record that the 'A.E.U. had taken up the matter with the Minister of Labour, who in due course in exercise of his powers under the

"Conditions of Employment and Arbitration Order 1305/40, as amended", directed that the matter be put to arbitration'. A hearing was fixed for Wednesday, 1 March 1944. The court under Justice Simonds reached the conclusion that the bonus in dispute was 'a gift from the company, and that the unions had no right to question its variation.' Unknown to the unions, George Pate announced at a board meeting a few days later that he had decided that Albion would be handing back to the Admiralty the management fee of £2,500 it had charged for torpedo production, less deductions of £885 for extra work the company had undertaken. This he felt was a 'nice gesture'. Ironically, too, the works received a visit from General Montgomery the following month. He toured every department, concluding by addressing a mass meeting of works and staff in order to convey the appreciation of the forces for the efforts being made on the home front. Although nothing appears to have been said, this must have rankled with the shop stewards and with many of the workforce who had received a slap in the face from the senior management for their patriotism, which had never been in doubt. Since 1939, the workforce had been contributing to a War Comforts Fund which had been used to send parcels of books, sweets and 'woollies' to the troops. This had subsequently been replaced by the sending of postal orders. By the war's end, 11,068 postal orders and 2,636 parcels involving an expenditure of £4,406 would be sent out. Pate's attitude was at best mean-spirited, and at worst, a misjudgement of almost treasonable proportions.

By the time the board met on 3 October 1944, the works had been on strike for five weeks. The ostensible reason for the strike was the dismissal of a fitter on 1 September. When this issue was resolved, the unions immediately substituted it with a demand for increased bonus and payment of a bonus in lieu to all non-bonus workers. Pate was 'quite adamant in his determination not to entertain the demands of the strikers'. Discussions at the meeting centred around the use of arbitration and a warning from Walter McFarlane that overheads were rapidly exceeding the levels that could be fully recovered from the government under the company's military contracts.

The strikers returned after an eight week absence, having failed to gain their objectives, and work was proceeding in an atmosphere of non-cooperation. The machine shops were going slowly, the shortage of output affecting assembly. A deputation from the 'workspeople' was brusquely informed by Pate that the bonus would not be raised and that those not entitled to it would receive none. By the end of November, there was still a level of non-cooperation. When H. W. Fulton returned in December, he found the factory demoralised in a way he had never seen before.

At a board meeting held at the end of February 1945 there was further discussion of the bonus scheme. The directors, augmented by Fulton and William Pate, managed to persuade their chairman that staff at Yoker whose service had been broken whilst in the employment of Coventry Gauge and Tool should be entitled to the long service award, to which Pate agreed. At the next meeting, he clarified that while their service should be regarded as continuous for qualification purposes, it should be discounted for purposes of payment calculation. This was followed by the board decision, by no means unanimous, that the bonus levels of sixpence and threepence should again be paid out. Thus, a matter that had soured industrial relations for three years was allowed to fester over the closing period of the war and the start of the reconstruction period, thanks to the intransigence of George Pate.

❧◈☙

Before the declaration of Victory in Europe on 8 May 1945, Albion submitted a programme of civilian vehicle production for approval by the Ministry of Supply, which included an order of 150 vehicles for South African Railways. It soon found it was able to sell everything that could be produced. Production was of course constrained by a number of factors: there was still a call-up of the 18-25 age group in operation, cooperation on the shop floor was reduced and suppliers were delivering slowly. The firm comforted itself in the knowledge that its rivals were also hamstrung by the prevailing circumstances. A.E.C. were rumoured to be producing 50 chassis per week, and Maudslay only 25. Word had been received from government that payment would be made in full for vehicle parts that were now surplus to requirement in view of the stoppage of military orders. The Clydesdale Bank had approached Albion with the offer of a £250,000 overdraft at three and a half per cent. Millburn Motors had also approached the company with a view to acting as its Scottish concessionaires, where sales arrangements did not already exist, and received a positive reply.

As if there were not more pressing concerns, George Pate continued to drag his feet on the bonus question. A board debate in November 1945 elicited the agreement that service in the forces was of course to be regarded as continuous so far as the long service award was concerned, but the chairman insisted the service period should not attract payment. At this stage Parkes, who had said little in earlier years, expressed himself unhappy. A decision was again deferred. The board meeting of 26 February was a crucial one. This time George Pate insisted that the strike which had been caused by the bonus issue should be looked upon as a break in service for purposes of long service award calculation. At this point he met with open hostility from the rest of the board, who proceeded to take decisive action outside the meeting.

The month which follows is not documented in detail, but Fulton records in his memoirs that the dispute had produced 'a head on collision' between George Pate and John Gray. While still at the Ministry of Supply, Fulton had learned that it was primarily regarded by senior staff there as being management's fault. Aside from who was to blame, the resultant strike 'had left behind an antagonism entirely foreign to the cooperative spirit' that had invariably applied at Scotstoun. While Fulton's fellow directors agreed with the point of view that he and William Pate held about the unreasonableness of George Pate's position, they had believed that as subordinates of the chairman and managing director, they were not in a position to do anything about it. Even if it were possible, they had felt that it would have been politically dangerous to attempt to overrule him.

However, the reluctant rebels were now ready to act. At the end of March 1946, George Pate declared at a board meeting that it would be 'preposterous' to make any further concessions in the bonus and long service arrangements. At an emergency meeting the next day, he advised his fellow directors, who appear to have conveyed their hostility towards him, that he had studied the company's Articles of Association and felt that since he was in the minority regarding the bonus arrangements, he was obliged to 'resile' from his position, and was prepared to do so. The rest of the board made it clear that this was not sufficient, and that they had now decided he was to be dismissed.

On 24 May 1946, at a specially convened board meeting, Pate, who had taken legal advice, announced that he was not accepting dismissal. One week later, a subsequent meeting elected H. W. Fulton chairman on a pro-tem basis, making it clear to Pate, who had demanded to be invited since he still considered himself chairman, that he was being dismissed for taking action 'detrimental to the interests of the company', refusing to take the instructions of the board and treating it with disrespect, and for approaching shareholders in an attempt to interfere with the board's workings. Pate reserved his position, passed on the view that the board's action had been 'mean and cowardly', and subsequently sued for wrongful dismissal. In due course he was awarded £60,000, which he gave to charity in a gesture that, compared with his attitude towards the workforce, was completely uncharacteristic. The board were prepared for this, had weighed up the consequences in advance and 'never saw reason to doubt' the wisdom of their action, since they felt that if he had carried on as before, he would have ruined the business.

In later years Fulton considered that Pate's decline related to his spell under the autocratic Beaverbrook, which he felt had resulted in the addition of 'a domineering trait' to 'an always somewhat ruthless character'. The combination, he felt, was 'undoubtedly powerful, but unsympathetic'. What had started out as a piece of small-minded penny-pinching by Pate had developed into a battle of wills, first with John Gray and the workforce, and then with the rest of the board. The magniloquent gesture with the damages cannot itself be interpreted unequivocally. It might either have been a genuinely charitable deed or just an expensive attempt at self-vindication.

The Albion directors, under Fulton's urgings, thus came to understand the disadvantages of corporate governance associated with the combination of the chairman's and sole managing director's positions. This approach tends to put too much power in one man's hands which, if unchecked, leads to abuses which have to be sorted out in public, with all the consequences for the company's good name.

In May 1946, with George Pate removed, the board's attention next turned to the appointment of an independent chairman. When T. B. Murray, N. O. Fulton, W. H. Fulton and J. F. Henderson had died, their estates included some £100,000 Albion ordinary shares each, the great majority of the Albion equity, which of course, they had deliberately accumulated in order to keep control. By 1946, all but Henderson's equity had been dispersed and much of it now belonged to private investors who had small holdings. Henderson's shares had been kept together in a trust and was the last large parcel in individual ownership. The board decided to invite no less than Jackson Millar, son of one of the original shareholders and recently involved in the Yoker pistol factory incident. He was asked to take up the post of non-executive chairman to represent the interests of non-directoral shareholders, and H. W. Fulton was confirmed as managing director. At last the company was able to give its undivided attention to the rebuilding of its business. Thanks to the damage caused by the industrial relations fiasco of the later war years, much had to be done inside the firm.

By June 1945 a new works journal had appeared, entitled the *Sunrise*, produced and sold by the Albion shop stewards with no management input whatsoever. This was not *ipso facto* a retrograde development, since one of its objectives was to provide a medium of communication between employees. Its front page declared that Albion was '100 Per Cent Trade Union' and that 'Union is Strength'. It seems to have been produced monthly and lasted until December 1947. It appears to have lacked widespread support on the shop floor compared, for example with the *Radiator*, and the vast majority of articles were written by trade union activists with hard hitting socialist opinions. *Sunrise* indicates that so far as its authors were concerned, there was a deep chasm between bosses and workers. The one was out to exploit the other, and the paper's first editorial condemned attacks in the national press on the work rate of the working classes. An article on the second page was subtitled 'All the Management Do About Mess — Mess About'. A number of pages dealt with the impending general election, and an article alleged that if the Tories were successful they would 'soon be monkeying with the people's savings' by making sure they would have none. It concluded that 'If you trust this gang with either your vote or your money savings, you are green enough to back Tommy Gardner in the St. Leger'.

While the magazine represented a legitimate point of view, one of the few non-activist contributions it received was from 'Analyst', in a letter to the editor:

> Dear Sir — Allow me to offer my congratulations on your having supplied a long-felt want in the Albion. The Sunrise has made quite an auspicious start. May I take the opportunity of placing what I hope will be constructive criticism? In the first place, is none of the shop stewards interested in any subject other than politics? I can assure you a very large percentage of your readers are interested in other spheres.

The editor replied that he had printed what had come in and was suitable, but that other kinds of contributions would be welcome. Unfortunately, they remained few and far between.

Articles on the 'Increase', 'What do You Know about Dispute Procedure?', 'The Students, the Poles, and the Miners' no doubt kept employees educated on socialist affairs and trade union issues, but are unlikely to have greatly entertained. Other pieces complained about the unrealistic targets set by ratefixers, excessive profits made by firms during the war (of which Albion was not a bad example), or gave a new episode of 'the Adventures of Angus MacSnoop' a 'boss' out to

8.12 Exhortation to boot Churchill out in Sunrise, 1945.

discover misdemeanours committed by shop floor workers. Without passing judgement on the left wing points of view expressed in *Sunrise*, there is no doubt that in its earlier years it had taken on a bitter flavour as a result of the disputes surrounding the bonus scheme.

If it undoubtedly reflected something of the poor relations between management and workforce, it also reflected a strong sense of identity with Albion as an institution which industrial strife was insufficient to destroy. H. W. Fulton, who regarded it a little extremely as 'vitriolic', was pleased to see its passing in 1947. It was understandable that he should do so, for by that year he had put in place a series of new measures designed to remove some of the circumstances from which *Sunrise* had emerged. By then he had introduced a number of reforms: a new pay award consisting of 50 per cent flat rate wage, with 50 per cent made up of bonus earnings, rather than 100 per cent, communications letters and meetings, regarded by the older directors as undignified and, last but not least, the qualification period for long service bonus was reduced from five years to two. *Sunrise*, in its final edition, was even complimentary:

The intimation made by the managing director at the canteen meetings regarding the reduction of the qualifying period for participation in Service Bonus from five to two years quite obviously "Rung a bell", to use his own phrase.

While the symptoms of a divided house at Albion were all too visible in *Sunrise*, its earlier editions undoubtedly also captured the mood of ordinary citizens on the eve of a crucial general election. The war had brought people of differing social backgrounds together and they had worked and lived together in a situation of common danger. They had accepted the constraints imposed on all by war and had been proud of what had been achieved. They assumed that when the war was over, they would share the rewards, particularly in the form of better housing and better social services. The troops, in particular, believed that this was what they were fighting for. In December 1942, the Beveridge Report commissioned by the coalition government, had dealt in clear and vigorous language with basic human rights, including the need for a comprehensive system of social insurance and the foundation of a national health service.

In the election campaign of 1945, Churchill and the leaders of the Conservative Party were less emphatic on questions of social reform, argued that it could not be afforded and warned of the dangers of socialism. It was inseparable from totalitarianism, they argued. Working people clearly saw that the desired improvements would not be forthcoming from the Conservatives nor from Churchill, notwithstanding his supreme leadership qualities. Labour pressed home its arguments. Its manifesto, *Let Us Face the Future*, promised widespread nationalisation, a national health service and a comprehensive social security system. The Conservative Campaign, which centred so much on Churchill being allowed to 'finish the job', was quite simply out of touch with the views of the nation as a whole. Labour won the contest with the largest majority it had ever gained in a general election, 146 seats, unsurpassed until 1997.

The problems faced by Attlee and his government were immense. The stock of national assets lost during the war was put at seven billion pounds. This was accounted for by loans run up, destruction and damage to property, shipping losses, depreciation and obsolescence of stock and the depletion of the nation's gold and dollar reserves. In addition, two thirds of Britain's export trade had been deliberately stopped and the economy had been completely reshaped to support the war effort. The number of people serving in the armed forces, civil defence and war industries had gone up four and a half times to nine million. Wartime inflation had been rampant, and the purchasing power of the pound had declined by one third over the period. Of these problems, the most pressing was that of exports. Britain had been forced after the cancellation by the U.S. of Lend-Lease to seek dollar loans, and to take less of these from the U.S. than she wanted. A Canadian loan was also negotiated to help bridge the gap. Massive currency repayments had to be made. International trade had to be rebuilt to help with the currency crisis, and the government set targets: in order to recover the position, the volume of exports would need to rise by at least 50 per cent above pre-war to repay overseas debts, and by 75 per cent if reserves were also to be restored.

Firms such as Albion were faced with a post war dilemma; whether to heed the government's exhortations and make a special drive for export, or to take the easier route of majoring in the home market, which was likely to be more

8.13 Visit of concessionnaires and official repairers to Albion, September 1948.

profitable. The board decided on the first course of action, and resolved to split deliveries 50/50 home and export. In 1946 all overseas managers were summoned to Scotstoun for their views on the state of their markets and future model requirements. Visits by Fulton to South Africa and Kenya and by Parkes to India, Persia and Iraq were undertaken to gain first hand experience. The directors gave exports special attention and conducted more overseas visits in 1947 and 1948. Representations were made to government to standardise vehicle widths. Overseas markets called for eight foot wide vehicles, British seven feet six. By the end of 1948, Albion was exporting 60 per cent of output. On its fiftieth birthday, at the 1950 annual general meeting, Jackson Millar was able to report that Albion had in the previous financial year 'exported the highest percentage of any commercial vehicle manufacturer in this country'. How had this been achieved? By concentrating on traditional, imperial markets certainly, but also by penetrating less promising territories in the middle east and Europe. Concessionaires had been appointed in Spain, Portugal, Uruguay and Syria.

The home market had also expanded, enabling Millar to claim in 1950 that home sales were also at their highest level ever. This was achieved against a background of continuing (but certainly reducing) parts and materials shortages, and the expected nationalisation of road haulage in 1947. Manufacturers did not like this, since it destroyed traditional relationships with customers. A further development not welcomed in the industry was the introduction in 1950 of purchase tax on domestic vehicles, which came in at the same time as tax increases

8.14 6 passenger station waggon, 1948.

on oil and petrol. These measures served to push vehicle users further in the direction of diesel engines.

The underlying alterations in models during these years were crucial to success. As ever, the export models introduced had to be more robust than for the home market, where the Salter legislation of 1933 still applied and roads were fully developed. Bonneted styles were still required abroad. Models to be sold overseas

Albion

CLYDESDALE

TRACTOR MODELS FT101 and FT102

WITH OIL ENGINE

OVERSEAS TYPE

Albion
SURE AS THE SUNRISE

ALBION MOTORS LIMITED

GLASGOW　　　　　　　　　　　　**SCOTLAND**

L 528F

BY APPOINTMENT
MOTOR LORRY MANUFACTURERS

8.15 Brochure for the Clydesdale tractor, FT 101 and 102, c 1950.

8.16 A SCWS Albion c 1950.

were often tested abroad while still at the experimental stage, as proof to export customers that Albion was seriously interested in exactly meeting their requirements. In 1947 the FT3AB Victor 31-passenger chassis was introduced for the export market, with a straight frame for higher ground clearance and the FT35 Clansman and FT37 Chieftain 5/6 tonners were also launched for export with new, light, Albion designed and manufactured diesel engines. The following year the CX41 Viking, also with a straight-framed passenger chassis was introduced for the export market, with a choice of two lengths. New models introduced to the home market included the FT5 and 7 light trucks and the slightly heavier FT17. In 1948 the FT101 and 102 Clydesdale diesel tractor units were introduced with the new light diesel engines. Numerous other model updates were made in the years up to 1950, although certain reliable units, such as engines, were not changed.

Part of Fulton's new approach to industrial relations involved invitations to the workforce to make suggestions regarding model names, and *Sunrise*, appropriately enough, records in 1947 that the names 'Chieftain', 'Clansman', 'Clydesdale' and 'Claymore' were the result of this initiative. Fulton had made two stipulations, that the names should be 'printable' and that they should have a Scottish flavour, and 610 suggestions were submitted by 300 employees. These would adorn Albion vehicles for decades to come. It has been suggested that some of these names were

8.17 An Albion demonstration truck in South Africa, c 1948.

'borrowed' by Albion from Halley, but any resemblance to Halley names arises from coincidence.

To support and rebuild domestic sales, a new depot had been opened at Norwich in 1946, and one followed at Lincoln in 1947. As post-war austerities began to lift, the Earls Court show was reinstated in 1948 and the Scottish Motor Exhibition revived in 1949. These developments symbolised a nation and an industry gradually beginning to regain full momentum. While there were temporary national economic setbacks during these years, particularly in 1947, the path was tending upwards. Profits rose from £137,486 in 1945 to £211,117 in 1946 and then to £248,013 in 1947 and £251,681 in 1948. Ordinary share capital was increased by £100,000 at this time to finance the continued expansion of business and replacement of assets. Profits in 1949 were £185,492 and £228,350 was made in 1950.

There were of course, board changes in the latter part of the post war period. Mr. Donaldson, long serving director in charge of purchasing, stood down from his post in 1948 but carried on as a board member, finally resigning in 1949. Walter McFarlane resigned as Company Secretary in 1950 but retained his seat, being succeeded as Secretary by J. E. Campbell of the accounting staff. W. C. Reid was appointed a director that year. At the 1950 board meeting, a bizarre and completely unjustified suggestion was made by a familiar but surprising visitor — that the board had been poor trustees of the shareholders' capital, and that they should be removed. Its author was none other than George Pate. The former chairman had recently been awarded his £60,000 in compensation for wrongful dismissal, and had the previous day delivered a circular to shareholders containing the arbiter's decision, which, he felt, vindicated his position. He went on to state that the £60,000 damages together with legal fees and unnecessary expenditure

146

on an employee pension scheme had cost the shareholders in excess of £200,000, and that the present board should be removed. Jackson Millar replied that the arbiter's decision did not contain the whole story, that Pate's attitude had begun to affect the company's relations with customers as well as having damaged relations with employees, and that the new pension scheme had paid great dividends in employee relations. The meeting swept aside Pate's motion and carried on with its business, and there the matter ended. It was omitted from the *Financial Times*' report of the A.G.M.

With its first half century behind it, the company again stood in excellent health on every front. An intimation of what lay ahead had already been given in 1945. In October that year, the chief executives of five commercial vehicle companies had met at the Dorchester Hotel in London and considered a 'scheme for fusion'. Albion, A.E.C., Dennis, Leyland and Thorneycroft had been represented, and agreed to recommend to their boards that a detailed plan should be prepared by Price Waterhouse, but it had been turned down by Albion and Leyland, at least. Nevertheless it was in the direction of combination that the future would lie.

CHAPTER NINE

The Leyland/Albion Era 1951–1970

By the turn of the 1950s, Albion had much to be grateful for: good levels of output, sales and profits, the war and its austerities beginning to recede into the distance and industrial relations on the mend. There were countervailing factors. William Pate observed at a board meeting in 1950 that productivity was running at only 75 per cent of pre-war levels and it was also noted that Bedford, Austin and Morris lorries were better priced than the lighter end of the Albion range. Another aspect of the threat emanating from this quarter related to the size of vehicle that these 'volume producers' now offered. Before the war firms such as Bedford, Ford and Commer only produced three and four ton trucks, but the Albion directors were now concerned that they had stepped into the heavy market and would be difficult if not impossible to match on price.

The problems thus posed were related to the segmentation of the road haulage market, where the nationalised long distance hauliers would pay a premium price for the extra reliability and robustness that an Albion vehicle provided, but where the private hauliers who were limited to short localised haulage work of up to 25 miles were very sensitive to price. At the same time there were good reasons why such businesses should still continue to buy Albions. The longer view of profitability was one factor that could be taken into account, but so was the existence of localised repair and service facilities. One notable business that had taken the latter view was the Scottish Cooperative Wholesale Society, which for decades had bought Albions. More recently it had purchased Morris, Bedford and Ford vehicles only to find that a combination of unreliable running and the absence of local repair and servicing facilities had made this an unwise choice. More Albions were bought.

It is interesting to speculate concerning the productivity drop perceived on the shop floor. *Sunrise* during its short existence in the 1940s painted a picture of a workforce always under tremendous pressure from management to reduce job times and work flat out. What was the truth? Undoubtedly William Pate would have statistics available to him on quantities of piece parts produced and their time value, which would put him in a position to judge with accuracy. No doubt, too, the removal of wartime pressures for production would have a psychological effect on all concerned, leading to a reduction in pressure for output. In addition to this, it is likely that the many shortages of materials and parts experienced over the post-war period would reduce factory efficiency and give rise to allowances which entitled the production workforce to bonus when strictly, it had not been

9.1 Truck assembly at Scotstoun, 1955.

earned. A reduction in incentive is also likely to have been provided as the new factories set up in the district by firms such as Euclid, Rolls Royce and Remington Rand recruited skilled workers by offering wages above the levels that Albion and the local shipbuilders were paying, which would sap morale. This, too, concerned the Albion directors, who had noted large defections of skilled men.

Against this background of mounting difficulties, they also discussed vehicle pricing policies. Export margins were falling, but it was not possible to raise the price and retain volume sales. One director had opined that Albion was 'selling too cheaply' to overseas markets, only to be informed that it was necessary to sell at list price minus 20 per cent. One short term possibility lay with the new designs that were in preparation for underfloor-engined chassis. It was felt that these would soon help overseas sales and margins. William Pate on these occasions again raised the possibility that if only one kind of model could be produced, then costs would reduce and profits would improve. History was repeating itself, and several of the directors remembered the discussions in a similar vein that had been held in the post-world war one period. The demands of the market had not allowed rationalisation to take place then, and Mr. Kemp the chief designer stated bluntly that even if it were possible to produce a truck at £200 less than the current cost, the market would not be there to sustain volume sales. The company thus found itself in something of a dilemma which, while it was far from being an emergency, caused long term concerns to which there was no obvious or immediate solution.

It was at this point of self-doubt that fate intervened. At a board meeting of 12 March 1951, Jackson Millar informed his co-directors that he had been approached by Henry Spurrier, Chairman of the Leyland company at the most recent commercial motor show regarding the possibility of amalgamation. The matter had only gone as far as an agreement that if either company wished to consider this step, they would speak to each other first. Before the board meeting, Spurrier had spoken to Millar to state that the Leyland board now wished to take the matter further, and that Albion was their first choice as a partner. Spurrier had made it clear that if Albion was unwilling, then Leyland would simply look elsewhere. Millar asked the Albion directors for their permission for him to investigate the matter further and report back at the earliest opportunity. He went on to offer the opinion that 'a fusion of the two companies would be beneficial to both, and...such a fusion would result in a strong and efficient organisation'. A lengthy and detailed debate ensued. The Albion directors, motor industry men of long standing, knew the Leyland Company well, and the following facts are likely to have been familiar to them as they discussed this crucial question.

Leyland started out in the 1890s as the Lancashire Steam Motor Company, making steam wagons and lawn mowers at its workshops in the village of Leyland, Lancashire, which was seven miles south of Preston. Although the engineering force behind the firm was James Sumner, the Spurrier family came into the firm to provide finance. The company changed its name to Leyland Motors in 1907, by which time it had a successful business making and selling delivery vans, dustcarts and fire engines. It did well during the first world war, when it made armoured cars and a variety of other military vehicles. After the war, the company raised more capital and like Albion, purchased a whole field of military vehicles for reconditioning and resale. In addition to this it launched a de luxe motor car, the Straight Eight, as a rival to the Rolls Royce. Further ill advised purchases of military vehicles were made at a point where the market fell, and the company found itself in cash flow difficulties and unable to persuade its bankers to help. It lost £755,000 in 1922 and a board shake up, instigated by the bank, ensued. The outcome of the crisis was a massive injection of Spurrier family cash and the arrival of Aylmer Liardet as a general manager. The Straight Eight was dropped and a successful line of cars and vans, known as the Trojan, was introduced, and the company continued selling its reconditioned vans. In 1924 it made a modest profit, and a recovery in the bus market soon afterwards helped it restore its fortunes. By 1929, the firm had paid off its debts and had returned to prosperity.

By this stage, Henry Spurrier III was prominent in its affairs, and it was trading extremely well in the bus business. Abortive amalgamation talks with A.E.C., which had a monopoly on the provision of motor buses in the London market, took place at several points in the late 1920s and early 1930s. During the rest of the 1930s, the company went from strength to strength, recording profits of £632,000 in 1937. During the second world war, it specialised in the production of tanks, bombs, shells and tank engines. It declined cost-plus contracts, which it felt would make it inefficient and sloppy. By this time its chief production engineer was Stanley Markland, whose sharpness and engineering competence had played a great part in its recent success. After the Dorchester Hotel meeting of 1945, Liardet, who was managing director and Henry Spurrier, who was general manager, recommended to the Leyland board that it should merge with Albion, A.E.C., Dennis, Leyland and Thorneycroft. The board turned this proposal down for reasons that are of great interest in the context of Albion. First, it was in principle opposed to the near-monopoly that would have arisen if the mergers had gone ahead, and it felt that customers would not have been at all happy about it. Second, it was only in favour of mergers that would leave Leyland in a dominant position or at least, not in a subordinate position, in the merged enterprise. In addition, it foresaw difficulties in such a merger of unity of control, which it felt would be impossible to achieve if diverse boards of diverse enterprises were combined. It is clear that by 1951, in the case of Albion, Leyland felt it could achieve the twin objectives of having a considerable if not dominant say in the running of a merged organisation, and that there was a strong chance of unity of control.

Also important to an understanding of subsequent developments was the rise, at Leyland, of Donald Stokes. Born in 1914, he had decided at the age of eleven that he wanted to join Leyland. He had been fascinated as a child by the company's buses. In 1945, having been identified as one of Leyland's rising stars, he had been asked by Spurrier to write a brief on the way Leyland should organise its export business in the post-war period. He recommended a simple structure with one person in charge reporting to the directors, and that the company should concentrate on sterling area countries which had not set up their own vehicle production facilities. He was put in charge of exporting in 1946 and immediately put his ideas into effect. There were tremendous advantages in Stokes' simple British Empire-based approach. English was spoken there, and relatively speaking, these seller's markets, with their lack of indigenous competition, were easier. Stokes operated his office on a blend of informality and strict efficiency. In the market place, he relied on projection of his own image with customers, always going for the 'top chaps' in an organisation. In due course he was hugely successful and responsible for massive export growth not only in the easier markets, but in places such as Denmark, Finland or Holland. From 1947, the company's sales rose at an average of 60 per cent per annum, and Stokes was bound for the board room. Quite clearly, he had an eye to reducing overseas competition from Albion through the proposed merger.

By 1949, Spurrier had been appointed managing director of Leyland, and to that extent can be assumed to have been the leading personality behind the approach to Albion, which by all accounts seems to have been made without any sense of urgency. Nor was it a time when mergers and takeovers were rife. The

1920s had been a period of corporatism, but the early 1950s was not. The merger was not proposed out of fashion.

Millar began the debate on the proposed merger by stating to the other board members that he and Fulton had been invited to discussions with the Leyland board, who saw substantial mutual advantage in the merger. William Pate replied by pointing out that the nationalisation of long distance haulage had undoubtedly caused Leyland to rely more on exports for growth and that it was probably after Albion's overseas markets. Although it was not known to the directors, this was indeed a significant factor in Leyland's thinking. Fulton replied that 'the real reason' was economy in manufacture. Albion, he felt, could not maintain the manufacture of a wide range of models at competitive prices because of its low volume base, and would need to restrict its activities to stay in business. Pate stated that he had always been of that opinion, but did not see why Albion needed to merge to rationalise its activities. Fulton replied that rationalisation would involve taking the huge risk of guessing what end of the range to concentrate on, and that merging with Leyland would enable a single enterprise to be able to present the entire range to markets that required it by producing one end of it on one site and the other end on another. Pate pointed out that a change of government could easily result in the re-privatisation of road haulage and a return to more buoyant market conditions which would thus remove the perceived need for a merger.

McFarlane, the former Company Secretary, after questioning the view that a merger would lead to more economic manufacture at Albion, then came round to the opinion that it would lead to a common pool of vehicles, with Albion concentrating on the light end of the range and Leyland on the heavy end. Reid of sales was a little more sceptical and felt that he would prefer Albion to stand on its own feet and that the merger would lead to Albion playing 'second fiddle' to Leyland. McFarlane voiced the view that questions of control were almost irrelevant, since the market already controlled Albion through competition. Reid was also concerned at 'staff adjustments', to which Millar replied that while these would undoubtedly take place over time, no redundancies were planned. Reid then stated that while, theoretically, fusion made sense, in practice he believed Albion could carry on alone. At this stage Fulton came in with the view that Albion must 'get down to a flow production' and that it could not be done alone. Competition from Commer, Bedford and Ford would be keener in the near future, he felt, and that meant flow production, but also that Albion could not risk range reduction on its own. Reid replied that his dearest concern was protection of the workforce, and that this might be a real issue since Leyland would dictate policy.

At this point Millar called for a decision. Did the board want to take the merger to the next stage or not? Pate said clearly that he felt Albion would be mistaken to go ahead. Kemp stated his disliking for the idea, but felt there was little alternative. The majority view was that the matter should proceed. Leyland's verbal offer had been to exchange five Albion shares for four new Leyland shares, and the financial fairness of that offer was therefore to be tested by both sets of auditors as the next step. Within a few weeks the auditors reported that their investigations had resulted in the conclusion that the offer was fair. In terms of the size of the two companies, Leyland was some two and a half times the size of Albion.

In his memoirs, Fulton disclosed the pros and cons of the merger as he saw them. Theoretically the means by which the merger would be accomplished gave Leyland control. Leyland would acquire all Albion's shares, becoming its sole owner, and Albion a wholly-owned subsidiary. The institutions and individuals which held Albion shares would surrender them for shareholdings in Leyland, which in total would amount to a minority stake. Fulton was convinced, though, that the benefits to shareholders and employees by way of greater prosperity and increased security more than offset the loss of control, but admitted that many at Albion were very sceptical and had to be persuaded. On 14 May 1951, a letter was sent out by the Albion board advising shareholders of Leyland's offer, with the recommendation that it be accepted. Full details were sent out in June, and at Albion's A.G.M. on 29 June 1951, Jackson Millar addressed shareholders in the following terms:

> There has, not unnaturally, been expressed in certain quarters regret that the sole remaining commercial vehicle factory in Scotland should be merged with an English company, but if this secures greater security for Albion, it seems to me that this is the proper thing to do. Your board would never have been party to any proposition had they not been completely satisfied that the Albion interest would be fully safeguarded. Albion will continue to produce Albion vehicles, and in cooperation with Leyland both of us should in time effect considerable economies in the administration of our respective firms, and should be able to produce still more and better vehicles in their respective field and still more economical prices.

In due course well in excess of 90 per cent of shareholders, who were mostly private individuals, accepted Leyland's offer and the merger, barring a few formalities, was accomplished. Effectively Leyland, through a share swap, had acquired a company valued at some £3 million without parting with one penny of cash. The transaction was completed without fuss or hostility and without the involvement of consultants or merchant banks.

There were several immediate outcomes of the merger. Two Albion directors were added to the main Leyland board, Jackson Millar and H. W. Fulton, while Henry Spurrier joined the Albion board, which he only attended intermittently at first. On 27 August 1951, Fulton tabled a note of confidential policy decisions that had been made between Henry Spurrier and himself. Albion was to continue manufacturing heavy vehicles for the time being. A twin axle, rear bogie vehicle capable of carrying 30 tons was to be jointly designed by Kemp of Albion and Tattersall of Leyland, to be built at either Leyland or Albion in due course. Leyland was most anxious that Albion re-enter the 30 cwt. to three ton petrol engined market again, even if, initially, vehicles had to be sold at uneconomic prices. Albion had recently stopped making such vehicles because of the competition from the mass producers. Quite obviously, as a first step, the two companies wanted to have a comprehensive range, and other moves towards rationalisation were to be left until later. Both firms were at this time working to capacity and their first concern was to satisfy their customers, but in any case, it had been agreed that the process of acclimatisation should be gradual. At a personal level, Fulton and Spurrier liked each other, and there grew up a 'friendly spirit' between the former rival companies.

As relations with Leyland developed, Albion carried on more or less as normal. Shortages of sheet steel were acute throughout British manufacture during these years, which caused accumulations of stock and called for increased funding for this extra working capital. The Clydesdale Bank obliged by raising Albion's overdraft limit from £500,000 to £750,000 in August of 1952. As far as Albion's overseas markets were concerned, the beginnings of protectionism could be detected. South Africa was a case in point. In 1949, Albion had planned to convert its sales and service depot, which was run by John L. Murray, T. B. Murray's son and Fulton's cousin, into a limited company. Government was already beginning to indicate that it wished to curtail its import expenditure and increase local employment, and Murray had advised on a trip to Scotstoun that Albion was well liked in South Africa because it had Scots rather than English roots, and that he in particular had very good connections with government. It was anticipated that the new firm could reasonably foresee an annual turnover of half a million pounds per annum, about one eighth of Albion's (then) annual sales. By early 1952, Albion despatches to Australia and New Zealand were beginning to be threatened by the introduction of import licences, at which stage Albion finally formed its South African subsidiary. Exports remained high however, although Albion found that to concentrate on these meant losing orders to competitors in the home market who were able to offer better delivery times.

As ever, minor model changes were made. These included the new Clydesdale tractor chassis FT1015 in 1951, and in 1953, the Claymore, which gave Albion a new model at the lighter end of the range capable of three or four ton payloads and helped achieve the wide vehicle range desired by the new combine. Further evidence of group thinking that year was the introduction of the FT111TP Scammell tractor chassis, designed to pull trailers designed by another firm shortly to be merged with Leyland, Scammell. During these years ancillary businesses were also being initiated to attempt to expand Albion's engine output. The Albion-Albatros marine oil engine was introduced in 1951 in conjunction with the Warwick Motor Engineering Company Ltd., but quite naturally did not lead to massive sales. In a similar vein, the Albion Cuthbertson Water Buffalo was launched in 1952. Developed by James A. Cuthbertson of Biggar (later awarded the O.B.E.), a friend of H. W. Fulton, this vehicle was a tracked, crawler tractor specially designed for work in bogland. It used the Albion 289H four-cylinder diesel engine and was manufactured in the North British Railway Engineering Works in Glasgow and subsequently in Cuthbertson's own workshops in Biggar. While the vehicle was highly successful for the purposes for which it was designed, it never sold in great volumes, although some were exported to Canada and an oilfield version with a larger engine was developed in the late 1960s. The vehicles had a very long operational life and at the time of writing, examples are known to survive.

By 1954, group policy was beginning to assume a much clearer form. That year, Fulton was able to advise the Albion board that the prime policy objectives were to eliminate the production of competing models at the Leyland, Albion and Scammell plants and to provide a complete range sufficient to meet group needs. He went on to confirm that the Claymore was selling at an uneconomic price and that in due course Albion would concentrate mainly on the Chieftain medium weight type of chassis. A new model was planned and would take about two years to bring out. Meantime, efforts would be made to ensure that Albion designs would not conflict with the Leyland Comet or Beaver chassis, an objective to be achieved by paying heed to the maximum tyre sizes for which vehicles would be designed. It was hoped that the main model to be produced by Albion would eventually sell in volumes of 100/150 per week. Also of note was the fact that Albion had started to produce a large 900 inch diesel engine for railway use, a project that had been initiated by Leyland. It had been decided in 1953 that the Albion HD chassis, designed as six and eight wheelers in 1950 for 12–14½ ton work, should be dropped since Leyland models covered this range. By the summer of 1954, concerns were beginning to be expressed that the dropped models were not being replaced by sales of the corresponding Leylands. At a more detailed level, piecemeal rationalisation was starting to take place. Leyland's Plympton (Devon) depot was converted to a joint Leyland/Albion site. Lincoln depot was closed down. Overseas, Mercedes Benz had received permission from the Indian government to manufacture light chassis locally, and Leyland the heavy chassis. Albion terminated its Indian concessionaire arrangements, looking to Leyland to provide outlets. Albion's Spanish and Argentinian agencies were merged with those of Leyland. By way of offsetting the tightening of overseas markets that was taking place, the U.K. government was increasing the availability of export credits, and both Albion and Leyland were beginning to look at these as ways to retain overseas business.

Meantime, there were significant board changes. Spurrier's increasing involvements outside Leyland had called for the strengthening of its senior management, and Donald Stokes was made a director together with Baybutt, the chief accountant and Pilkington, who was in charge of technical and design. At this stage H. W. Fulton was made Leyland's deputy managing director, with W. West an assistant managing director. The friendly nature of the merger, if it required confirmation, received it with this appointment. When Spurrier attended Albion's board meetings, his comments were always reasoned and there was little on show of the autocratic streak for which he was renowned within the combine. Matters inside the group appeared to be moving as rationally and as harmoniously as they could.

Inside Albion, there were board changes too. Since his accession as Albion's managing director after the war, Fulton had adopted a policy of fitting jobs to

155

men, rather than the reverse. He had asked each director in turn to nominate a successor in the unlikely event he was to be 'run over by a bus'. Reid and McFarlane had been appointed on this basis, and in 1951, W. P. Kirkwood, Pate's nominee, was made works manager when he stood down. There were other interventions: J. E. Campbell, who had succeeded McFarlane as Company Secretary in 1950, died in 1952 and was replaced by James Maxwell, formerly the company's cost accountant. In 1954, M. J. Camplin of Sales, W. P. Kirkwood works manager and A. Craig Macdonald, production controller, were appointed directors. Macdonald was responsible for both production control and buying. In short, the policy was one of internal succession, although the standard Albion pattern of having a director for works, sales, design, technical and buying was subject to little variation. This resulted in a constant engineering predominance on the board with finance, arguably, under-represented. Moreover, this was an area where Fulton, on his own admission, was not strong. On the positive side, the policy meant that a relatively high representation of graduates, albeit with an engineering bias, were selected, which the Albion directors always believed gave the board a greater ability in problem-solving than the standard non-graduate approach of motor firms such as Austin or Morris. Those who stood down from their front line posts, usually on reaching retirement age, were invariably available as non-executive directors. If this policy regarding directors bred loyalty and enabled career paths within the company to be mapped out in advance, it also posed the twin dangers of lack of fresh thinking and an over-emphasis on engineering considerations. Nevertheless, it could be argued that a company that had not only survived but prospered for more than half a century must have been doing something right.

It could also be seen clearly that by 1954/55, the merger was proving a success. In 1946, 3,400 Albion employees had produced 1,500 chassis. In 1951, 3,250 employees produced 2,800 chassis and in 1955 3,523 employees produced 3,565 chassis, which gave a profit of £579,128 before tax. The fruits of business and engineering competence and group collaboration were undeniable. So too were the difficulties both the company and the industry generally had to face during these years from the fluctuating transport and economic policies of successive governments. Each time there was an anticipated change of direction in these areas, buyers would stop purchasing until the position became clear and boards and designers would hold their breath in case new designs and ranges were rendered obsolete. This happened in 1953 as rumours that the conservative government was about to denationalise transport broke out. This came about when the Road Transport Act was passed that year, which removed the 25-mile limit on haulage business for private hauliers. A Road Haulage Disposals Board was set up to sell the nationalised British Road Services fleet, but this was only partially done and about 16,000 vehicles were retained in public ownership in five companies dealing with general haulage, special traffics, contracts, parcels and meat. Sales were affected by the vehicles sold off at this time. Nevertheless, this did not alter the fact that these were prosperous years generally for British manufacturing industry, who were calling for greater volumes of goods than ever before to be carried by road. This was a firm foundation on which commercial vehicle manufacturers could build.

It had been made clear to Fulton in 1954 that as group deputy managing director, he had full responsibility for the group's engineering and technological policy. Parts of this were to his liking, such as Leyland's gradual incorporation and uptake of Albion-designed gear boxes for their chassis, but the Albion board had been less enthusiastic about Leyland's intention to design a joint cab for all group vehicles. Fulton and Pate both agreed this was not wise, since different styles and sizes of chassis required different cab detailing, but less contentious matters such as the rationalisation of models and major units soon took up his attention.

In this respect, 1955 proved to be a watershed year. Coinciding with Spurrier's knighthood, the process of rationalisation and centralisation within the group appears to have accelerated. At the beginning of that year, Leyland Motors in Wellington, New Zealand took over responsibility for Albion sales and servicing, and at home the Albion Hull depot took on responsibility for Leyland's customers. In May 1955, there was a request that Leyland and Albion set up a common costing system. Albion calculated product costs from the earliest days, and it is not known what the request led to, although a change of accounts classification appears a possibility. In June of 1955, all Leyland, Albion and Scammell export sales were united together in a common operation based in London, where each company in any case had sales offices. Each firm retained responsibility for home sales, however, which meant that at this stage, Albion still took home sales orders in London, Scotstoun or at any of its depots, and salesmen were still employed in all these locations. Credit control was to be centralised at Leyland. In October 1955, Albion was asked to provide capital and revenue budgets. Discussion at the Albion board on the assumptions which were to underpin these budgets seem to suggest that these had never been prepared before at Albion. Transfers within the group of components were taking place at cost up to this point, but these were henceforward to be given a mark up of five to seven and a half per cent in order to simulate arms length trading conditions within the group.

In February 1956, Fulton informed the Albion board that as a result of the reorganisation of duties at Leyland, he no longer required to act as deputy managing director, and similarly, Sir Henry Spurrier no longer acted as deputy managing director of Albion. Clearly, this was a greater loss of position to Fulton than to Spurrier, but the underlying rationale for this reversal is not clear. At the same meeting Fulton informed the Albion directors that 'the question of policy on the type of vehicle to be produced at Albion was still being pursued'. This enigmatic statement was not clarified, but hindsight would reveal that it masked a series of difficulties that Fulton was facing in connection with the group policy for Albion. Meantime, the tide of rationalisation continued. Six monthly financial accounts were being compared with budget at Albion and Leyland board meetings, and while Spurrier intermittently attended the Albion board, two new directors were added in September 1956: Donald Stokes and

Stanley Markland. At the same meeting Spurrier indicated that a full merger of sales activity should be contemplated in the future. A group cost reduction committee was to be set up to discuss ways of reducing costs. Further afield, Albion and Leyland's sales and servicing operations in Australia were merged, while in Birmingham, Albion's depot was sold in order to replace it with a joint Leyland/Albion depot.

Much of this rationalisation was entirely predictable and exactly in line with the plans laid at the outset of the merger. The financial tightening, too, could only be regarded as laudable in principle and a reflection of the relatively low profile given in the past to finance at Albion and the rise of Baybutt to the Leyland board. On the technical side, however, integration had reached a level that Fulton was finding it difficult to stand. Initially he was glad to discover that there was little overlap between the ranges provided by Albion and by Leyland. The latter's strength lay in the heavier double deck and single deck bus markets and in heavy goods vehicles, while Albion was good at medium-weight trucks and tailor made heavy models, especially for overseas markets. Fulton's greatest difficulty lay with Leyland's intention to redesign its cab, on which it had already expended a great deal of time and money. While it was logical that this should be made to fit the requirements of other group companies, Fulton concluded that Leyland's design did not meet the requirements of Albion customers.

As well as disagreeing on the cab, Fulton had difficulty in persuading the Leyland Design Policy Coordinating Committee that a new epicyclic-geared hub reduction rear axle designed at Albion should be developed and made standard. In spite of successful tests, the committee was reluctant, although he eventually gained his point. Similar difficulties were experienced on the subject of engines. The committee wanted to change from a four to a six cylinder engine for the revised Chieftain vehicle, and Leyland built a prototype six cylinder unit which was preferred to Albion four cylinder designs. Fulton could not agree with the extent to which these and more minor proposals by Albion were being overturned, and he discussed this with Jackson (now Sir Jackson) Millar. Millar advised him to accept the majority decisions of the Leyland board in these cases, but as a matter of conscience, he could not. He saw the continuation of such a dismissive attitude as 'the end of the tradition in which I had been raised' and the loss of significance of Albion as designers.

Accordingly, as a matter of principle, he informed the Albion board on 1 April 1957 that he intended to resign from his two board positions, effective 30 April. Jackson Millar replied that he was 'greatly distressed' at Fulton's decision. At this stage Fulton retired from the meeting, whereupon Millar informed the board that Leyland was equally sorry about it. Kemp at this stage stated his view that Fulton's departure would cause 'considerable uneasiness throughout the staff' and particularly among the senior members. In his ten years as managing director, Kemp noted, he had gained the confidence and respect of all works and staff employees. Kemp felt it would be highly desirable that Fulton remained on the Albion board, even if as a non-executive. The rest of the Albion directors were in

9.2 Albion at the Scottish Motor Show, 1957.

agreement, but nothing was done and Fulton left the company. Millar was, however, able to reassure the board that Sir Henry Spurrier had stated that Fulton's resignation would not affect production or employment at Albion. Thus resigned from Albion a solomonic figure in the mould of his father, having restored industrial relations harmony after the war and fought Albion's case at Leyland for six years.

Fulton was later at pains to stress in his memoirs that there was no animosity whatever in his relations with those at Leyland, and that his resignation was tendered out of principle alone. In subsequent years he looked back and wondered if 'he had done the right thing', especially since what he fought for had eventually been implemented. Both Leyland and Albion ended up using the hub reduction axle, the Leyland six cylinder engine gave problems and was not introduced, and the four cylinder model was reinstated. In due course, the Albion cab rather than the Leyland became the standard one. From a personal point of view, his resignation was perhaps to be regretted, but he concluded that if he had given in on these points, the Albion board that he left behind might not have been in existence. Writing in 1964, Fulton was still of the opinion that his advocacy of the merger was broadly correct, but that his views on its rightness were not quite so strong as they had been in 1951.

9.3 Albion Claymore with box van body, mid 1950s.

Fulton had other interests which had developed during his Albion years. He had become a director of Scottish Aviation in 1954, and had been closely involved with the North British Locomotive company in connection with the Water Buffalo project. He took up a directorship there on leaving Albion. A final speculation regarding H. W. Fulton relative to the Leyland merger is perhaps appropriate. It seems likely that part of the attraction of the merger for Fulton was the opportunity it gave him to widen his horizons and operate on a larger stage. While this would not have superseded in his priorities the protection of Albion's interests, it would have been less than human of him not to think of this aspect, which may in a small way have helped bring about the amalgamation.

Fulton's successor at Albion was no less than Stanley Markland. It had been observed that during these years, Leyland was effectively run by three men, Spurrier, Stokes and Markland. The latter was in charge of production, and was regarded as a 'brilliant' production engineer with a feel for everything he did and a 'marvellous eye'. He was an authoritarian of the old school, occasionally had a rough tongue, and was alleged not to be comfortable in the company of others of similar ability. His independence of mind brought him into conflict with Spurrier on numerous occasions, and as a result, he felt a certain deep seated insecurity in his relations with his boss. Over against this, he nourished an ambition to succeed Spurrier as managing director of Leyland.

On succeeding Fulton in 1957, Markland announced at the Albion board that each executive director was to provide him with a written monthly report which he would combine into a report for the chairman of Leyland. He felt that Albion's

9.4 A Caledonian on the Albion stand at the 1957 Scottish Motor Show.

immediate priority was to get the new Chieftain vehicle into production within its budgeted cost. In common with best industry practice (no doubt emanating from Ford), the new chassis had been engineered for a budgeted cost that would enable it to sell for £200 less than the present model and be able to compete with the mass producers. He was concerned at Albion's overhead recovery rate, which had risen recently from 280 to 308 per cent, and expressed the hope that with volume sales, it could be reduced to 250 per cent. The present Albions were expensive and lacked an attractive appearance. This impressive command of management accounting and the latest model costing techniques was new to Albion. At a subsequent board meeting, Markland announced that he had noticed a 'general slackness' throughout the works. At around the same time, he introduced quarterly budget comparisons and cash flow forecasts and supported the results of a systems study across the group that resulted in the conclusion of Baybutt, who had led it, that there were numerous redundant forms, statistics and procedures in use at Scotstoun that could be eliminated. If this happened, staff could be reduced by ten per cent. Markland was certainly living up to his reputation.

Over against the imperative to reduce costs and improve efficiencies in the middle 1950s, there were contrary pressures. The Claymore vehicle, while it kept

161

9.5 H. W. Fulton and J. L. Murray on an A6 Albion outside the Kelvin Hall, 1957.

a light option in the Leyland range, was at first riddled with engine problems, and never sold at good prices or in viable numbers. External factors such as the Suez crisis caused fuel shortages and chopped chassis sales to only 2,520 units in 1957, bringing about factory layoffs. Government, as ever, had played its predictable part in altering the parameters of the industry's operation: road tax on vehicles was increased in 1955 and at the beginning of 1957, vehicles in excess of three tons were at last permitted to exceed 30 miles per hour. 1958 introduced the threat of re-nationalisation and purchase tax and the raising of HP deposits, which slashed sales to 2,256 units. Social conditions were also imposing extra costs. At the end of 1955, Albion had conceded an extra week's holiday for staff, which was becoming standard in the area. Competition, too, was tightening. Visits by senior sales staff to continental motor factories and shows confirmed that Mercedes Benz, Man, Henschel and D.A.F. were already a force to be reckoned with. When William Pate and Francis Henderson retired in the spring of 1957, they picked an opportune moment.

In model terms, 1958 was a watershed year. The long-awaited basic 7-ton Chieftain model, created for volume sales and incorporating the latest technology, was at last launched. In a fanfare of publicity both in the press and at depots, it

9.6 FT 39 Albion Victor bodied as a pantechnicon at Brisbane, late 1950s.

made its appearance complete with its new Albion-designed cab and hub reduction axle, the very things over which Fulton had resigned. Later that year, a sister vehicle, the Clydesdale, was launched for 9 ton loads, with the same features, which were also applied to the heavier Reiver. The details of how these reversals of engineering policy were achieved have not survived, but the innovations in question formed the basis of Albions for many years. Of special significance was the cab. It represented a leap forward in design when compared with the rectilinear lines of previous Albion cabs. Cabs, if desired, could of course be provided as standard at the factory, before bodybuilding. The new model's curving all-steel roof, front panels and windscreen gave a very up-to-date appearance and more important, all round vision. By today's tastes, it perhaps seems a little ugly, but it imparted to the vehicles to which it was fitted a sense of lean, no-nonsense bulldog-like ruggedness. The cab was known as the LAD cab, the initials standing for Leyland, Albion and Dodge. An arrangement had somehow been made for the three companies to share the tooling costs, and it was produced by Motor Panels of Coventry, the presswork specialists. It was in production until the early 1970s. A revised Claymore was also launched in 1958.

The year brought some sad developments. Sir Jackson Millar died, leading to Sir Henry Spurrier becoming chairman of Albion. Millar, who had also been a director of the nearby Yarrow's shipyard, was a stabilising and moderating influence within and without the company, and had helped bring Scammell lorries into the group. Meantime, at Albion's board meetings, Craig Macdonald was beginning to exercise his powerful personality, and in due course he would come to be referred to at Leyland as 'the Claymore'.

The steps forward in model improvement had come too late in the year to save the company from its worst post-war results to date. It recorded a loss of £96,000

163

9.7 Albion Victor VT 17N supplied to Ceylon Transport Board, 1959.

9.8 Albion Victor VT 17 N for Boston Transport, Barbados, 1959.

on sales of the 2,256 vehicles, the year having been affected by difficulties with competition and exchange control problems in foreign markets, as well as UK governmental manoeuvres. That year there were deflationary measures which included a high but falling bank rate, purchase tax was altered and there was again a rumour that road haulage would be renationalised. In the era of Sputnik three, the genesis of the E.E.C. and the commencement of the Forth Road Bridge, it seemed that no one cared for the truck industry. Thereafter, however, matters improved. In 1959, purchase tax was removed from commercial vehicles, interest rates fell, and with the re-election of a conservative government at the end of 1958, the threat of renationalisation was removed. The new models proved immensely popular, and from 1959 were fitted with the new four-cylinder diesel engine advocated by Fulton. Albion turned a profit of £230,000 in 1959 on sales of £9 million, and blipped its accelerator in readiness for the booming 1960s.

At this stage, there were highly significant changes afoot not far away at Bathgate in West Lothian. Here, a giant motor factory was about to begin production. The British Motor Corporation (B.M.C.), formed in 1952 by the merger of the Austin and Morris motor companies, had been persuaded by government that expansion in Scotland was the way forward. This was part of a regional policy approach in which dispersal of the generally booming motor industry was seen as an engine of growth and job creation in economically blighted areas. The West Lothian district had suffered badly from the demise of coal and shale mining, which had left a number of symbolic tombstones in the shape of the many 'bings' that dotted its environs. With massive government financial assistance, the new complex was completed in 1960. Its four factory buildings extended to 1,250,000 square feet. One of these housed Europe's largest machine shop, one an agricultural tractor factory capable of producing 500 units a week, with a third and fourth capable of assembling 1,000 trucks per week in both built up and knocked-down (kit) form. The trucks concerned had in some cases been produced previously at B.M.C.'s Longbridge plant, from which some of the machine tools had been sourced.

Inside the factory, B.M.C.'s Boxer, Laird and Mastiff trucks were produced, together with light vans. There was a certain amount of overlap with the lighter end of the Albion range, but Bathgate's output was mostly of the 'biscuit tin lorry' variety, although a bonneted vehicle, the WF, sold abroad. Production was organised around moving assembly lines, and after initial bedding-in problems with green labour, the factory would soon employ between four and five thousand people and would occasionally touch 80 per cent of its output capacity. It was hoped both by the government and B.M.C. that the plant would increase its considerable multiplier effect in the local economy by encouraging component suppliers to locate nearby. Within a few years, a factory already planned for Rootes at Linwood would be completed on a similar scale, this time producing motor cars and employing 7,000 people. Hopes were high that these plants would dynamise the Scottish economy.

9.9 Albion Caledonian tanker, 1960.

Meantime, at the top of Leyland, Spurrier, Stokes and Markland were beginning to widen their horizons. Running a very successful truck and bus empire was one thing, but Stokes was by the late 1950s beginning to see advantages in certain overseas markets in being able to offer comprehensive ranges of vehicles of all kinds, cars included. By 1960, Leyland was seriously eyeing up the Standard-Triumph Company with a view to cooperation in Sweden and New Zealand and began conversations that year with Alick Dick its chief executive.

166

9.10 LAD cabbed Chieftain in 1958.

Dick's hidden agenda was to sell the company in its entirety, since he felt that it was too small to meet the mammoth retooling costs associated with new models of car and did not really have the volume base to recover them. By this time, its financial position was deteriorating. Leyland soon came round to the view it should be acquired. From 1961, when Leyland eventually took over Standard, its problems were emerging and thereafter took up an increasing amount of the senior executives' time, especially Markland, who ran it. Just afterwards, fresh problems occurred. Leyland in 1962 soon saw B.M.C.'s attempted acquisition of A.C.V., which owned A.E.C., Leyland's arch rival in the bus market, as a major threat. As this reared its head, no less than Chrysler of the United States made a tentative approach to Leyland with a view to acquiring it. This was anathema to Spurrier, who disliked the idea of ownership of his company passing outside Britain, especially in the light of adverse experiences he had had with United States firms. The threat acted as a goad to Leyland, and by June 1962, Leyland had acquired A.C.V. to gain extra mass and strength. In addition, the takeover would also enable Leyland to eliminate A.E.C.'s competition abroad, particularly in Africa. Here, Leyland and Albion had combined their interests by forming Leyland South Africa Ltd. and had in recent times been forced into heavy discounting and losses by A.E.C. This was eliminated at a stroke.

It follows that from the early 1960s onwards, Leyland had moved into a different league and that the vastly increased scope of its business and takeover activities must of necessity have diverted the attention of its principal directors

9.11 Unidentified Albion chassis at Scotstoun, 1958.

away from Albion. In many ways the Scots firm had overcome the major hurdles that stood in its way and was ready for increased prosperity and a reduced level of scrutiny. 1960 was a year of records. Chassis sales were 4,700 and in addition 365 engines, 6,238 gearboxes and 1,331 rear axles were produced. This took the firm's cumulative chassis output total to 103,396. After tax profits of £647,000 were made and expansion was needed. Property was purchased at the corner of South Street and Balmoral Street, and a new machine shop was built. All this was achieved with 2,258 works and 800 staff employees. The production of 4,212 chassis the following year generated £658,000 in after-tax profits. The market boom in 1963 and 1964 took output even higher with after tax profits rising to £829,000 that year. 1963 chassis output was 4,885 and in 1964 it was 6,102. That year the market was so buoyant that Albion temporarily stopped accepting orders. At the end of 1963, Sir Henry Spurrier stood down from his business posts, gripped by terminal illness. To the deep disappointment of Markland, who had in the meantime presided over the rehabilitation of Standard, Spurrier selected William (Bill) Black of A.C.V. as his successor at Leyland. Black was 70 and Markland 59, and the latter resigned all his posts in disgust, leaving a vacancy at Albion for a chairman. This was filled by Donald Stokes. In 1964, too, W. C. Reid retired from the Albion board.

In 1965, two Leyland board members, Trevor Webster and Dr. Albert Fogg, both engineers by background, were added to the Albion board. By this time the rationalisation of Albion production had resulted in the commencement of a £2 million extension to the west of Albion's main office block to provide a 750 foot long moving track for chassis assembly. When this was completed in 1966, it may have been the longest of its kind in Europe. At last flow production in a key area of the factory had been achieved. Chassis could begin as a frame at one end and drive away at the other. with the help of this new technology, output rose from 5,736 chassis in 1966 to 7,463 in 1970, almost 150 per week. What were these chassis? They were predominantly a blend of the hugely popular Chieftain, Clydesdale and Reiver trucks. From 1964 onwards, some of these were supplied with the new luxurious Sankey-built Ergomatic tilt cab, designed for use by all group companies and still looking remarkably modern today. In 1961, Albion had re-entered the double-deck bus field with the 72/74 seater low-height Lowlander chassis and the single-deck market with the 41 seater Viking VKHL. These and updates sold in modest numbers throughout the 1960s. In 1967 a tractor chassis capable of hauling a Scammell trailer was added to the range.

The moving track involved a period of acclimatisation and it took a little time before factory component output rose to its demands. The culmination of this process was the acquisition in 1969 of the former Harland and Wolff shipyard premises across South Street and directly opposite the main works. Here axles for Albion and her sister plants in the group were to be made. Management needed strengthened during this unprecedentedly prosperous period in Albion's history, which quantitatively speaking, was the most successful in its lifetime. In typical Albion fashion, internal appointments were made. W. P. Kirkwood had retired in 1966, and Jim Pollock,

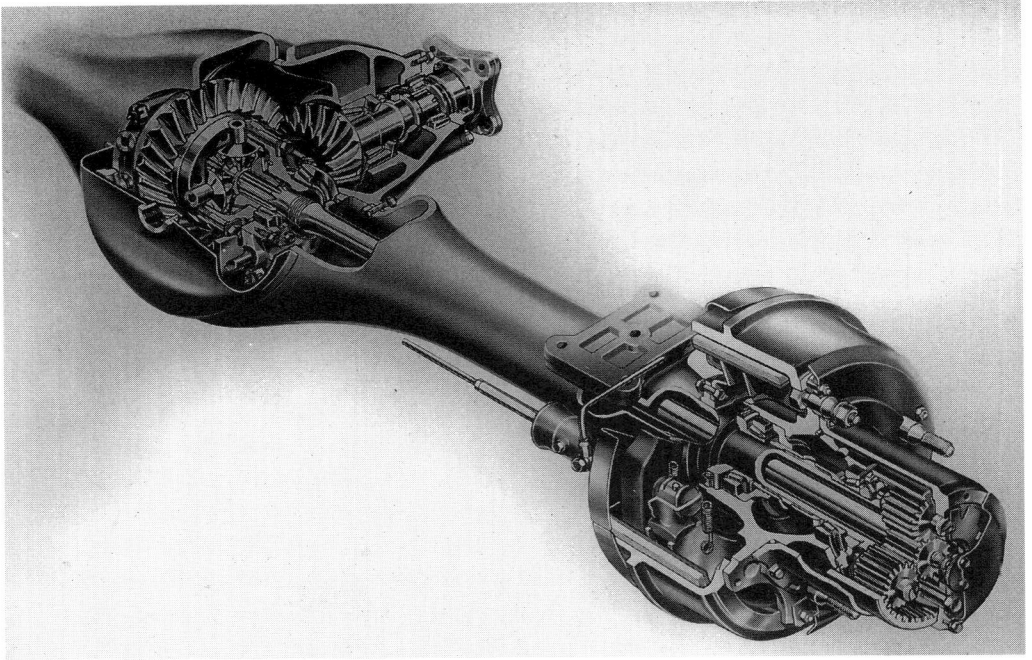

9.12 Double reduction rear axle, as fitted to the Chieftain and Victor range from 1960.

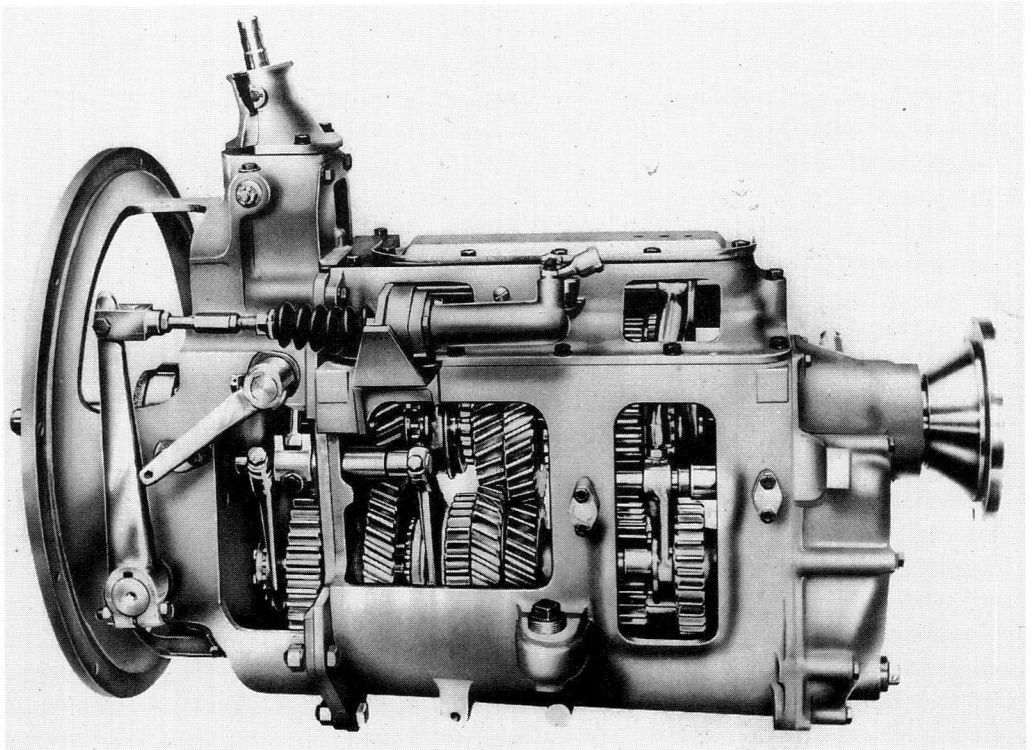

9.13 9 speed gearbox Type GB 248, optional on Clydesdale and Reiver models from 1960s.

9.14 Super Clydesdale diesel tractor with Ergomatic tilt cab, late 1960s.

an Albion apprentice and graduate engineer, was appointed in his place as director and works manager. Ron Dougal, who was commercial manager, was also added to the board and in 1969, J. Milloy, engineering manager, was appointed a director. In 1969 profits after tax were £1.3 million on sales of £20.7 million.

In the corporation as a whole, massive and unprecedented change had taken place by the end of the decade. Encouraged by Labour Prime Minister Harold Wilson and Anthony Wedgwood Benn, the Minister of Technology, Leyland by this time under the chairmanship of Stokes had taken over British Motor Holdings, the parent company of B.M.C. in late 1968. The takeover had also been encouraged enthusiastically by the Industrial Reorganisation Corporation set up by the government in 1966. It had been felt by the government that the British motor car industry was far too diffuse, produced too many models leading it to inadequate profits and investment and radically needed rationalised. Over-manning of a very substantial order was suspected by the government. Indeed, the industry was already showing signs of ailing.

The two corporations had discussed merger as early as 1964. B.M.H. felt that Leyland's profitable medium to heavy truck business would splendidly complement

9.15 West Gate, Scotstoun works, late 1960s.

its high volume, low margin light truck operations, based at Bathgate. Leyland's motives for the merger were complex and often less than clear. The expansion of truck operations and the elimination of competition abroad were obvious advantages. The vulnerability of Leyland to takeover made the leap in size following the merger a desirable outcome. Aside from that there were personal ambitions and corporate aspirations. After many negotiations over several years, a deal was done and most of the British-owned motor industry thus came to be part of British Leyland Motor Corporation. The huge combine operated on some 60 sites and employed 188,000 people at its outset. In 1968 it had made over one million vehicles generating sales of £907 million and profit after tax of £20.3 million. In 1969 the figures were £970 million and £20.8 million respectively, and some 50 per cent of output was exported. Towards the closure of the merger, Leyland's expectations of B.M.H. began to reduce as its profit forecasts were somewhat suspiciously revised downwards, but the national interest seems to have played its part in convincing Leyland's board. Stokes was sure that B.M.H. could be sorted out and that the new corporation could be successfully rationalised.

This would have grave implications for the future of Albion, which was shortly to become part of B.L.'s Truck and Bus Division. Soon its fortunes would be linked with the nearby Bathgate factory of B.M.C., its proximity to which could not be ignored.

CHAPTER TEN

Sunset Meets Sunrise: Emasculation and Decline Under British Leyland

The last legally filed accounts for Albion Motors Ltd. were for the year 1970. Incredibly, in an administrative step taken out of convenience, the Albion figures for that year were aggregated with those of Leyland (Scotland) which was described in the notes to the accounts as a 'branch' of Albion! An outsider reading these accounts would be led to the conclusion that Albion had taken over Leyland (Scotland), as the Bathgate factory was now known. If anything, the reverse was in due course to prove nearer the truth. Under British Leyland, the destinies of the two plants were to be drawn together.

The figures in themselves are revealing. Albion's net fixed assets in 1969 had stood at £1.75 million. The combined figure for 1970 was £8 million. Albion's stock, which had always included consignment stock at overseas depots, branches and distributors, had been £7.6 million in 1969, whereas the combined figure for 1970 was £17.8 million. The profit figures revealed no such proportional relationships: Albion's after tax figure for 1969 had been £1.3 million. The combined figure for 1970 was just short of £1.5 million. Little wonder that Harriman of B.M.H. had courted a relationship with Leyland. His truck profits were marginal. That notwithstanding, realities inside the new Truck and Bus Division eventually shifted power eastwards.

Legally speaking, it had been decided that Albion should cease trading operations as of 1 October 1970 and this was recorded at Companies House. In future, its results would be consolidated as part of the Leyland Truck and Bus Division, at Leyland. By 1970, the Albion board consisted of Jim Clegg, long serving accountant at Albion and latterly Company Secretary, Ron Ellis, Truck and Bus Division board member and sometime sales executive under Stokes, (by now Lord Stokes), together with Dougal, Milloy and Pollock. Macdonald had by this time gone. The earliest board minutes under British Leyland from 1970 disclose that Albion was selling a mixture of trucks and components to some 30 associated companies within B.L. worldwide as well as to its own external customers. Since the mid 1950s, the export proportion of sales had dropped to about 30 per cent, representing a tightening of exchange control and import regulations abroad that was never again to slacken. Arrangements were being made to transfer all Truck and Bus Division spares business through a special depot at Chorley, and centralised engineering and sales functions were being planned at Truck and Bus Divisional offices in Lancashire.

At the Albion board meetings the Leyland influence meant that finance now featured much more strongly than in the 1950s. Debtors were regularly reviewed, and budget performance monitored. Gerry Wright the Truck and Bus Division finance director attended from time to time. Albion's profit for 1970 was £3.8 million before tax on sales of £24.7 million.

By 1971, no doubt under the advice of Ellis, R. J. (Jack) Smart was appointed a director of Albion. Smart was based at Bathgate, where he was General Manager. Milloy had taken up a new post as transmission engineer at the Divisional Engineering office in Leyland. He was replaced on the board the same day by Bob Beresford, Chief Engineer at Bathgate, to coordinate the engineering functions at both plants. At the board meeting where this took place, Ellis was critical of Albion for having lost production due to a short strike and overtime ban by clerical workers in pursuit of a pay claim, asking Dougal, who was General Manager at Albion to tie up with Smart to see if Bathgate could help with the backlog. It appears unlikely that Dougal, a forceful personality, would have liked this.

By mid-1971 it had been decided that the LAD and Ergomatic cabs in use at Albion should be replaced by the 'G Range' cab made at Bathgate. In 1968, a new cab facility had been added to the Bathgate complex and it made no sense that the ranges of vehicles in the two different centres should have different cabs. There were, however, to be detailed differences to reflect the premier status of the Albion vehicle, including a superior trim. The cab, of handsome appearance, was to be launched at the Earls Court Show of 1972, and was of the 'tilt' type. Unfortunately, it would say 'Leyland' on the front, but the words 'Chieftain', 'Clydesdale' or 'Reiver' would denote the vehicle's origin.

The year 1972 had been a poor one in the annals of labour relations in the United Kingdom. The number of working days lost through strikes had gone up steadily from about three million in 1965 to about 24 million in 1972. The miners had gone on strike on 9 January, and by 9 February, large scale power cuts had caused a state of emergency to be declared. One and a half million workers were at one point laid off work at that stage as power supplies fell. It was 25 February before the miners voted to return to work, having been awarded a 17–20 per cent increase. Rail strikes and 'work to rule' campaigns disrupted life further. By the end of 1972 the conservative government imposed a total wage freeze, which was to be gradually relaxed over the next few years. Notwithstanding these great difficulties, which had caused some layoffs and some disputes at Albion, the situation was kept under reasonable control and clearly, although details have gone, good profits had been made. Short term loans to the division stood at £5.25 million. Once again, Albion's moderate industrial relations heritage had stood it in good stead, although there is no doubt this was coupled with competent personnel management.

By 1973, a 'Scottish Operations' set up had been put in place for management accounting at Albion. Utilising Bathgate's detailed budgeting expertise, a system of weekly budget reporting was created for regular management reviews to be held at the plant, where a small management accounting team was appointed. Albion had of course put costing in place at the very beginning of its existence, but this was a new refinement. Managers of each of the four main shops, Machine

10.1 Signs of the times — Albion exhibits as part of B.L. at the 1969 Scottish Motor Show.

Shop, Gear Shop, Axle Shop and Chassis Assembly each received weekly reports on hours clocked analysed into productive and non-productive elements, units produced, efficiency ratios, manpower levels and ratios, overtime cost and hours, absenteeism, together with scrap costs and manpower. Each was compared with budget weekly and monthly. In addition, incentive bonus performance by area was monitored against budget. At some time prior to the B.L. era indirect workers had been consolidated into the scheme, which removed much of the subjectivity over indirect bonus that had proved so contentious in the past. These innovations and the retirement of Dougal at the end of 1973, followed by Pollock the next year, distanced the plant further from its historical heritage.

By the mid 1970s, Albion was a modern, well run and profitable plant, but now fully integrated into British Leyland. Financial figures from this period, if they survived, would be meaningless: there were variations in transfer pricing policy over the period, spares income was taken elsewhere in the division and product engineering and marketing decisions were taken off site. Sales made to sister companies were at transfer price. The profit figure was only of internal validity relative to budgets. In 1974, Geoff Richardson, Bathgate finance director, took over as Albion plant director and after a stay of one year, returned to Bathgate. He was succeeded by Jim McCluskey, formerly of Bathgate, and then by Andy Barr who had been hired from Chrysler, Linwood as production manager. Each of these, after short spells as Plant Director, moved back to Bathgate to promoted posts. From now on, the senior managers at Albion would form the local board without directoral titles. At times, long serving Albion employees noticed a difference in style and manner to that of the old Albion directors. A certain confrontational approach germane to the industry as a whole but not Albion was sometimes introduced and did not go down well with middle management, but several of these short stay directors demonstrated a single-mindedness that had at times been lacking in the past. By 1977, Roy Dale, a mellower figure with substantial overseas experience was in place at Albion, calming ruffled feathers.

That year, a telling incident is said to have occurred. An old Albion customer ordering 60 buses for Barbados at a price of about £1 million insisted on having the famous oval saltire plaque that had featured on earlier Albions provided on each bus. Apparently, the Minister of Communications and Works personally demanded these, and they were duly supplied. This was not the only case where this type of thing happened. Good names die hard. By 1978, Albion was still producing some 170 trucks per week plus 300 gearboxes and 300 axles. Employment stood at just under 3,000.

As for the gigantic B.L. combine of which Truck and Bus Division and Albion plant were part, its performance to 1973 had shown disappointing progress. Employees had risen to 204,000 and output to 1,161,000 units. The highest profit achieved to date, £27.9 million for 1973 after taxation, was insufficient to re-invest in the rationalisation and new models required. In 1974, the use of oil as a political weapon by O.P.E.C. had brought inflation to 17.1 per cent and caused the three day week, which lasted for the best part of three months. This took B.L. into a loss of £6.7 million after tax, causing net cash assets of £51 million to swing to borrowings of £86 million. After a meeting between Tony Benn and Lord Stokes, the Ryder Report was published in March 1975. It was accepted and became the basis for the rescue of British Leyland by the government. Funding of £700 million, staged over the years 1975 to 1978 was promised, plus overdraft facilities of £200 million. To 1981, the government committed £1,200 million, to be provided in loans and equity, which meant that the company was de facto nationalised. Progress in industrial relations and efficiency improvements was set as a condition for the support, but this was abandoned in due course. Growing dissatisfaction led to the appointment in late 1977 of Michael Edwardes as chairman and chief executive.

Just prior to this, an internal reorganistion of Truck and Bus Division into Medium/ Light Vehicle Division and Heavy Vehicle Division under Des Pitcher had caused the decentralisation of certain elements of Leyland-based divisional management to be dispersed to an office (bizarrely) situated in the not very prestigious Wester Hailes district of Edinburgh. This led to an influx of highly paid financial analysts into Wester Hailes and then, subsequently, into Bathgate Plant. The timing and public relations functions and elements of marketing were also decentralised at this point. It was widely felt at the plants that the exotic salaries and relocation packages involved were not the best possible use of public money and that the change did nothing to solve the division's or B.L.'s fundamental problems. Nevertheless, there were some communication benefits from the devolution of divisional functions to plant level.

After Edwardes came into Leyland, what had been the Truck and Bus Division was further restructured into Trucks, Bus, Parts and Self Changing Gears under David Andrews and Ron Hancock, as part of a devolved structure in which management were expected to exercise their 'right to manage' in an assault on productivity and industrial relations. Just shortly before these developments, a sad act of great symbolic significance took place in Scotstoun: the destruction of the Albion main office block, ostensibly in the name of providing improved loading and unloading access, but perhaps also to mark the parent company's determination to turn its back on the past.

Of more significance was what had happened at Leyland. The company had chosen to invest massively in a new £33 million state-of-the-art assembly hall for heavy trucks at its home base. The whole range was in need of a revamp, and heavy vehicles were to be the first to be completely redesigned. A magnificent truck facility was erected complete with fully computerised stores in which to produce the new heavy range. The undernoted statistics indicate what subsequently happened to Leyland's truck business over the period and beyond.

B.L. Commercial Vehicle Statistics

000s	1977	1978	1979	1980	1981	1982
Production	120	131	125	130	83	89
Home Sales	61	67	70	60	42	42
Home Market 000s	230	260	304	272	220	226
Home Market Share %	26.5	25.7	22.7	22.0	19.1	18.6
Import Penetration %	16.5	21.8	23.0	24.1	31.4	29.9

While the above figures include light vans, the pattern is clear enough. The home market fell dramatically between 1979 and 1981. Inside this the market for heavy trucks fell by 44 per cent between these two years. The reasons are connected with the unprecedented strength of the pound which both cheapened imports and affected exports, a devastating double edged sword.

Under Edwardes, facilities were cut and rationalised. What this meant was that in June 1980, Albion ceased to assemble vehicles and the facility was transferred to Bathgate, which had abandoned agricultural tractor manufacture. The unthinkable (which most had feared would happen) had come to pass. Albion naturally saw substantial redundancies, but also witnessed partial reversal of the tide as the axle facilities of Leyland, Guy and Alford and Alder were moved to the plant, followed by the lighter axles from Bathgate. Edwardes was typically outspoken about the circumstances in which trucks had to be sold, lambasting government exchange rate, interest rate and tariff policies which had of course played their part in the collapse.

By the time of Edwardes' standing down from B.L. in September 1982, a £990 million support package over a two year period had been pledged to the company, but downwards alterations in the trucks plan had again to be made. Both the market and Leyland commercials were again in free fall. In 1985, out of 266,000 commercials produced in Britain, B.L.'s share was 90,000, of which only 15,000 were trucks, and by 1986, the total output was 228,000, of which B.L.'s share was 69,000 and only 11,000 were trucks. Ten years earlier, Albion and Bathgate alone had been producing four times this amount between them. From the early 1980s, it was widely expected that further rationalisation would happen and in 1984 truck production at Bathgate ceased. Successors to the Bathgate and Albion models had been progressively introduced at Leyland from 1980–81. At the level of objective rationality, it was perhaps logical that these should move into the new assembly plant at Leyland, the most modern facility within the truck division. However, the centripetal force of head office in times of retrenchment is well understood in Scotland, which has so often seen the loss of jobs to the south in recession, and the transfer was in some quarters read as a denuding of Scottish facilities to protect the centre.

It was therefore in suspicious circumstances that the completely redesigned successors to the old 'Redline' (Bathgate range) and 'Blueline' (Albion range) were built at Leyland. Unanswered questions remain as to why the company decided in the late 1970s that it would be the Leyland-produced heavy end of the range that would be updated first. It was this decision that seems, with hindsight, least rational of all. The low volumes of vehicles affected by the redesign would have given a low return on the massive investment in the new assembly hall even if the original estimates of volume on which it would be based were realised, which they could not in the circumstances have been. Given that the market collapse was particularly savage in this range of vehicles, the new assembly hall must have appeared a white elephant, crying out to be filled. Whatever the reasons, it ended up producing both the volume end of the range and the heavier end of the range, and the key was finally turned in the lock at Bathgate in 1986. The site was soon flattened and returned to its pre 1960 state, having gone from grass to grass in a quarter of a century, and West Lothian's job famine resumed acute proportions once again. For those involved, the stress was barely tolerable.

The effects of this radical remarshalling and downsizing of resources was manifested in 1983 at Albion, now run by former personnel manager Jim McGroarty, where there was a dispute over voluntary redundancy involving what one writer has referred to as 'relatively trivial issues'. In reality it must have been a manifestation of the cumulative stress inevitably involved at times of retrenchment, which it is surely thoughtless to trivialise. By the mid 1980s, only 1,100 men out of some 3,000 employed in 1978 remained. At the end of 1987, the North Works at Albion was closed, ending 85 years of automotive manufacture on the site. The new focus of manufacture was the South Works, the former axle factory. It was planned in 1987 that this site, which represented about half of the previous floor space, should build axles of light and heavy capacities in volumes of 50,000 per annum, with substantial reductions in fixed costs.

Very soon there were developments at Leyland. The market in 1987 was at its lowest level since 1952. In February Leyland Trucks was sold to the Dutch company D.A.F. The conservative government, by then flat out on its programme of privatisation and keen to be rid of the remaining pieces of B.L., wrote off £300 million in debt. Incredibly, the market fell still further in 1988. 40,000 trucks had been sold in the U.K. in 1987, falling to 37,000 in 1988. Ford, of course, always seen as a benchmark competitor by B.L., had itself suffered abominably. Of 1,138 Ford truck franchises operating in Britain in 1981, only 577 remained by 1988. The Ford truck business had in 1986 been merged with IVECO, owned by Fiat, with managerial control passing to IVECO. The market had indeed been hit by a series of body blows from which even the best were unable to recover. The final step in the abasement of Leyland D.A.F. came in 1992 when the company, weakened by losses caused by the further collapse of the U.K. truck market in the recession, fell into receivership. At this point the Albion workforce, which had been downsized to 750 by 1990, was reduced to 530.

What had happened to Leyland's truck business? The decline of British Leyland as a totality and the industry in general has been frequently analysed, the broad reasons for its demise are widely accepted and many apply to trucks. The product range was always too large, leading to expensive and unprofitable production, although that was being rationalised. Quality was suspect, in some plants more than others. Marketing had been poor but was improving. There remained the problem of underinvestment, labour relations were often difficult and there was overmanning, although all of this was being tackled. Added to this, there had been weak management, again improving.

It would be completely wrong in the circumstances to omit to blame successive governments for policies that in a period of about 20 years swung from seeing the industry as an engine of export-led growth and job creation to viewing it as a thorn in the flesh to be disposed of and passed to overseas control at a knockdown price. It was also harmed by unrealistic exchange rate maintenance strategies that reduced exports and eased imports. This altered market conditions, rendered planning null and void and demoralised all concerned. The catalogue of problems is not pleasant to read although these were being tackled by the late 1970s, and it should be noted that foreign-owned manufacturers based in Britain often did little better than B.L.

It is not the intention of this study to supply a full and detailed analysis of the reasons for the decline of B.L.'s truck business. For this to be attempted with more precision than has already been provided, a full examination of a whole number of factors would need to be undertaken. These would include, for example, the trend to larger trucks and the fairly recent legislation that has brought it about, together with the development and effects of the roll on, roll off ferry; containerisation and its effects and the rise in Britain of foreign-owned indigenous plants such as those of Volvo. Also worthy of consideration would be the seemingly inexplicable desire of Britons in the past few decades to prefer foreign made goods in general and to suck in volumes of them detrimental to the balance of trade which, in relation to motor vehicles, has become chronically negative. A further major factor requiring study would be the investment and strategic decisions of those in management and government responsible for B.L. Truck and Bus between 1974 and 1985. How sound were their model and investment decisions? To what degree were they overshadowed and de-prioritised by the urgent need to save the cars business and the launch of the all important Metro and Maestro vehicles? How much cash was siphoned away from Trucks to cars pre-1974? In the course of any future attempt to answer the above questions in detail, some points of great Scottish relevance would surely emerge. What part, for instance, did 'branch factory syndrome' play in the disinvestment of Leyland Trucks from Scotland? In the absence of evidence, final judgement must be reserved even if suspicions lurk.

***

If the foregoing study of Albion puts some parts of the jigsaw puzzle that is the history of the truck industry in place, it surely also has a wider value as a case study in management, particularly for the period when Albion was independent, or relatively so. Any summary of a firm's management would need to include an examination of its interaction with technology, marketing and sales, finance and the man-management dimension, as well as a review of the firm's top management and strategic behaviour. The information that has survived does not enable each of these to be examined systematically or fully, but a broad picture most certainly does emerge.

So far as technology is concerned, earlier chapters have identified a range of traits. First, a preference in the eyes of Murray and Fulton for reliability and gradual evolution at a time when practice during the first ten years of the century was very diverse and change rapid, which of course led the founders towards commercials, where a level of reliability at least equal to the horse had to be achieved. Their predilections were therefore reconcilable with demand in the market place. The second phase, post 1918, saw the founders step back in this regard and allow younger subordinates to take the engineering forward into live rear axles and away from chain drive, and into the era of pneumatic tyres and four and six cylinder engines. Thereafter, the continued tendency of Albion to over-engineer and over-specify was noted, but this was combined with a greater sensitivity to the market place and developments in the industry, which saw the firm enter the world of 'overtype' cabs and six and eight-wheel vehicles.

The Salter decisions of the early 1930s saw a new era, and with the new regulations and tax regimes affecting vehicle design, the firm acquired very considerable expertise in the use of alloys and special steels and the ability it had to maximise vehicle loads relative to weight, thus eliminating over-specification. This was manifested in its wheel springing systems and later, in the development of its epicyclic-style axles. Its slightly late entry into diesel engine design and manufacture was not down to technical reluctance, but rather to an ill judged commercial decision that hinged on a wrong assessment of future trends, relative to excise tax and this was soon reversed. Underfloor and rear engined vehicles followed fashion as did tilt cabs. In all this, the firm retained its commitment to high quality through sound, long term vehicle testing, and the evidence that survives suggests that remedial work on Albion vehicles after sale was relatively low throughout. None of the above, in spite of cost penalties to Albion, harmed the customer, whose loyalty to the firm became a mainstay of the business. The use, too, of engineering graduates and the existence of training schemes and a culture of sound engineering helped the company, as did the purchase of the best tools and test equipment, at every turn. In the long term, even the costly aspects of this approach to technology redounded to Albion's good: it earned customer loyalty although at times prices were lower elsewhere. Over the piece, therefore, Albion's management of technology was on balance positive, healthy and good.

Albion's performance in sales and marketing was on a par with other leading manufacturers throughout. It had a varied range of outlets from about the first decade of its existence, when concessionaires were extensively used, then this was progressively changed to regionally-based offices, showrooms and depots in the U.K., and abroad, a mixture of depots and concessionaires was instituted. The records surrounding overseas sales are few and fragmentary, but information survives for the year 1920 and this is likely to have been typical both of the pre-war and post-war scene. Some 330 out of 1,506 chassis sold were sent abroad that year to overseas customers, about 20 per cent. 158 of those exported were sold to customers in Madeira (7), India (20), Straits Settlements (31), Java and Sumatra (11), India (28), Burma (4) and Baghdad (47) through London purchasing agents or confirming houses.

In this, the presence of a London office must have been absolutely crucial. A direct sale to 'the colonies' of 32 vehicles was also made by the Crown Agents' London office, with the balance going directly to customers in all parts of the world via local companies in Denmark (1), South America (10), Cuba (27), India and Ceylon (38), New Zealand (18) and 'Miscellaneous' areas (14). Five were sent to the Albion Motor company of Australia, while Chapman, a local concessionaire, took 54 vehicles. The Albion Motor Company of Canada took five. In due course the Albion subsidiaries in Australia and South Africa would assume a life of their own in the post world war two period. While Albion sales executives and directors undertook extensive sales tours at various points in the company's history, the impression is given that there were ready markets for the company's vehicles in the empire and beyond and that Albion was far from being the only U.K. seller to these countries. This is not a fault, and it appears to have done everything possible to encourage export sales in an era that predated intensive air travel.

All of the domestic sales activities outlined above were based on the Scottish and English motor shows that from the earliest years became a focal point for the industry's customers, and where new products tended to be launched. Persistent evidence comes from the Albion records that very few sales were actually concluded at these shows, but leads were obtained and information given out that often led in due course to new business. Then as now, the shows were a locus of social contact within the trade as well as a serious marketing medium. Here, too, information on the latest engineering trends would be gained.

In addition to the above, publicity brochures using advertising messages that appear even today to be surprisingly modern were used extensively, as was advertising in the quality newspapers and trade press. Of note was the distinctive quality-related messages the company constantly emitted, and the rising sun emblem with later verbal embellishments. These were of course devices similar to those used by many of the major companies of the day.

Mention must also be made of the Albion style. The impression is given that in common with a few other companies, Albion attempted to lift the company's and industry's image by employing sales staff with qualities of 'gentlemanliness', which initially coincided with the aspirational lifestyles of Murray, Henderson and to a lesser extent the two Fultons. This would perhaps only be true of a few of the motor manufacturing companies. The portraits of the London managers, especially Parkes but to some extent Young, give off an up-market image, no doubt connected with the graduate calibre of these gentlemen, those they would hire and their successors. This would reinforce the 'quality' image that Albion wished to portray.

The usual indicator of the effectiveness of the 'marketing mix' employed is market share. There are difficulties in this regard with Albion. As a car and commercial producer to 1910, a producer of commercials and small vans to the late 1920s and then as a producer of medium and heavy trucks and a smaller amount of buses thereafter, it would be difficult to know precisely which domestic or export market figures to compare output with over time. Added to this problem is the fact that Albion's output of engines, gearboxes and axles ebbed and flowed individually from 1951 with changes in Leyland sourcing policy. Albion, in short, moved in different markets over the period. In the circumstances it is difficult to think of a more satisfactory index to marketing success than Albion's constantly rising output figures, although this ignores relative performance. There is, though, no reason to doubt that Albion's marketing methodology, which changed over time, was anything less than adequate. It may frequently have been more than that. For example, in the 1930s, its share of UK sales rose from two point four per cent of commercial vehicles (in 1930) to five point three per cent (in 1938), and its outstanding export performance in the 1940s and 1950s also deserves mention.

So far as human resources are concerned, the story outlined is one of success, given the circumstances in which Albion has operated. Situated in 'the Workshop of the British Empire', cheek by jowl with the steelworks, shipyards and engineering works of Glasgow, it had no great problems in recruiting the skills it required from the talented and mature pool of labour available. There were occasional difficulties, though, as the stresses of wartime depleted the workforce, or when the new post-war factories sought to recruit by paying superior wages. Industrial relations at Albion, with few qualifications, has been a success story, notwithstanding the firm's close proximity to, and close involvement with 'Red Clydeside'. A harmonious atmosphere, as was demonstrated, has usually prevailed, the exception being when intransigence on either side has been shown, as happened towards the end of and slightly later than world war two, when Albion became a 'closed shop'. A key element in this phenomenon was the wisdom of Fulton senior and junior, mirrored in later periods by the skill of, say, Jim Pollock. When stresses and strains did show, they were frequently related to the vexed question of bonus working which is widely regarded as open to abuse, divisive and not conducive either to proper motivation or to proper management. Albion's industrial relations restraint was well tested in the 1970s, at a time when the general motor industry experience was a music hall joke. By and large, the centre held and there were few strikes, a credit to all concerned.

On the staff side, too, Albion cultivated a climate which encouraged graduate engineers into key posts which, as was noted, was most unlike many of the car companies, and this underpinned the technical and general management competence of the company for many years. As for the rest of the staff, appropriately educated and qualified labour was easily acquired from the workforce living in and around Glasgow, and there were seldom difficulties.

As Appendix Two shows, Albion's finances, apart from the depression years of 1921 and 1931–1933 were strong and on the whole, well managed. There were seldom problems with cash flow, and the Clydesdale Bank's constant support was only used intermittently. At the end of 1920, as the refurbishment of military vehicles took place, the Clydesdale loaned Albion £240,000. At the same stage, the company had investments of £175,000 which could have been sold to achieve virtually the same effect. By 1921, a year when a loss was made, the overdraft had dropped to £124,000 and the stock had fallen, thanks to the sale of the investments. Thereafter, the bank was seldom used, wartime being the main exception. During world war two, changes of government plans for types of vehicles to be produced and supply shortages caused stock buildups and a need for assistance with liquidity, and the bank was more extensively used at this stage, but of course all orders were for the government and eventual liquidity was guaranteed. The same had applied during world war one. Outside war periods, current assets were always well in excess of current liabilities.

The basis for Albion's strong liquidity position was its profitable trading, which only lapsed into losses (small ones at that) when the manufacturing sector in general was in difficulties thanks to the macro-economic situation. At a time of low, single figure interest rates, Albion's after tax returns, as Appendix One shows, were invariably handsome. This deteriorated slightly in the 1940s, when the firm's decision to pursue export business resulted in reduced margins in markets that have been described as 'easy' but where there was competition among U.K.

manufacturers sufficient to make companies such as Leyland look at mergers with A.E.C. and Albion in order to eliminate it. It must be noted, too, that Albion's most successful years both in terms of vehicles sold and profit made were the 1960s, when it expanded under the wing of Leyland Motors. It is in the area of profitability, though, that Albion is perhaps deserving of some criticism. As an engineering-oriented firm, it tended to over specify especially prior to 1932. While it tended to be able to recover its costs in seller's markets, it paid a penalty when the shoe was on the other foot. Albion 'priced a design' rather than 'designing for a price'. This changed when Leyland took over, and vehicles were designed to a cost budget with a selling price and margin in mind. It is likely that Leyland adopted this practice from the example of Ford, who had been involved in it since the 1920s. This goes some way towards explaining the improved margins of the 1950s and 1960s.

The blame for these earlier approaches must be taken by the Albion board, who never accorded finance the respect it deserved. Although figures were regularly discussed, costs and margins were not extensively debated, and the cost office, although it operated from the earliest years, appears to have been regarded as a routine function, operating on the basis that it calculated margins achieved *ex post facto*. The model proliferation of the 1930s caused by Salter was a case in point. There appears to have been no discussion of costs and margins in a situation that, granted, was something of an emergency. Serious consideration of budgets, too, was only introduced during the Leyland era in an atmosphere where the engineer was king.

The beneficial aspect of this engineering bias related to reinvestment. As the figures show, there was a consistent and substantial reinvestment of post-tax profits, seldom much below 50 per cent and frequently well above it, as in the middle 1920s or the late 1940s. Even after the periods of loss-making, when shareholders had been relatively under-rewarded, as in 1930 to 1933, there was a strong urge to reinvest as the figures for the later 1930s show. The upshot was a constant replacement of machine tools and equipment, although much also flowed into working capital to expand sales from time to time. This underpinned the quality and reliability of the products.

Capital gearing has been commented on frequently in earlier chapters, especially relative to the desire of the first directors to retain control. Gearing was only ever an issue between 1918 and 1931, when the firm had debentures outstanding, taken out to finance post-war expansion. Their effect on the gearing ratio dwindled thereafter until they disappeared. On balance, what deficiencies there may have been in Albion's financial management proved to be minor ones, taken in the context of the company's other strengths.

In terms of the quality of Albion's management, a number of issues arise. The functional basis of its organisational structure appears to have given few problems although a suspicion creeps through that certain departments, such as the drawing office in the post-world war one period might have been isolationist and pushing for policies that made life easier for it rather than better for the company. This may indeed explain some of the personnel changes that were made during the George Pate era, when intransigence would not have been tolerated, and might also explain why there was such widespread support for, and frequent discussion of, the possibility of standardising models. The dwindling health of Murray at this stage makes this scenario more credible, but in any case, it never became a major problem.

The Albion experience in man management reinforces several common sense principles, the first being that the lack of mutual respect between those who govern businesses and those who make the products on which they depend leads to conflict and potentially, disaster. This was exemplified in the era during and after world war two, when industrial relations suffered. By contrast, the Fulton era exemplifies how important to peace and harmony an appropriate rapport between managers and managed is. The current foolish practice of explaining away the behaviour of contentious, self-interested and aggressive senior management as a question of 'style' is made all the more ridiculous when Albion's experience is examined in detail. The firm's history also provides some useful material for reflection relative to corporate governance. There were seldom problems when Albion was run by compatible personalities related by blood or marriage, although the other directors, often promotees of these men, emerge from the board minutes as competent but dependent and perhaps reluctant to make radical suggestions or speak their mind. When the original directors moved on to be replaced by a single figure with the responsibilities of both managing director and chairman, the outcome, for a time, was not to the company's benefit. The firm learned the lesson, early, that these positions were better split, predating Cadbury by more than 40 years. It must be said, though, that this brush with governance issues was an isolated episode.

At a strategic level, the Albion case provides food for thought. There can be no doubt that measured either by longevity, reputation, financial or technical performance the firm may be considered a success, even a great one. The question arises, however, as to whether or not it achieved its full potential. During the period when the Fulton brothers, Murray and Henderson were in their prime, it did exceptionally well, but its growth was inhibited by the four leading directors' desire to retain control. Given that control was a dominant impulse in their thinking and that it gave them the satisfaction, income and lifestyle that they wished, it would be inappropriate to insist that they should have widened the capital base and proceeded to 'conquer the world'. The depressions of the early 1920s and 1930s would possibly have made that difficult, but major strides forward were taken in the post-Lacre period from 1910 to 1914.

There seems to have been an opportunity for radical growth towards the latter part of the 1930s as markets boomed, but the urge was not there. From this point onwards, there was no aspiration on the board for substantially wider horizons, and the absence of an outstanding entrepreneurial talent among its number thereafter restricted it to the notion that at Scotstoun it had reached a plateau of success that was more or less sufficient for all concerned. Even H. W. Fulton, whose elevation to managing director was in the circumstances a breath of fresh air, lacked the large scale sense of adventure that could have led to a major strategic acquisition that would have left Albion in the driving seat or that would have produced a major expansion or diversification. Again, there were too many canny, empirically grounded BScs on the board for this to have been allowed to happen. Hindsight suggests that if it had taken place, subsequent events, including the Leyland takeover (for that is what it was) might have turned out quite differently.

When Leyland took over B.M.H. in 1968, it clearly made an error of judgement, and as Lord Stokes later mused, it might have been 'much more sensible to let it go broke and then bought some of the pieces'. This too might have left Albion in better shape and Scotland with a larger slice of the motor industry than she now has.

There still remains the question of whether Albion, coming from where it was in 1950, could have survived until today as an independent vehicle producer if she had said 'no' to the Leyland merger. Leyland averred that if Albion had said 'no', then it would have gone away, which was consistent with the English firm's stated desire to achieve unity of control. There is no guarantee, however, that others within a decade or two would not have made overtures to Albion. The crucial point was that at the time of the merger, Albion's shares were all in private investors' hands, and a hostile bid could have been mounted and succeeded. This makes the 'either/or' picture of Albion together with Leyland, or completely independent, something of an artificial comparison.

Allowing that no such alternative takeover took place, it follows that Albion would have had to survive the turbulent conditions that lay ahead: the market collapses and foreign invasions of the late 1970s and throughout the 1980s plus the takeover culture of that period. It is not impossible that Albion could have survived as a vehicle producer, but it *is* impossible to say.

CHAPTER ELEVEN

After the Eclipse: The Rise of Albion Automotive

'Awake, Albion awake!'
(William Blake, *Milton*)

The insolvency of Leyland DAF in 1992 was the darkest moment in Albion's history. At this stage it did not know whether or not it had a future. In fact, it later emerged that the collapsed DAF parent company had intended to close it in 1994 as part of its recovery plans. If Albion workers were in the dark, so too were employees of the complex of vehicle manufacturing and components plants at Leyland in Lancashire that had continued to produce trucks after Albion and Bathgate had stopped making them in the 1980s. While the arrangements for administrative receivership enabled all the former Leyland DAF businesses to continue trading for the time being, potential buyers of trucks would doubtless be wary of acquiring vehicles from a company that might not exist in the future, for this would mean that there was no spares or service backup. Production and sales did continue, though, and thus Albion's major customer remained in business.

What was to follow was a series of rescue operations that would result in the retention of every major plant in the former Leyland DAF group of companies, albeit with much reduced manpower. This was achieved through management 'buy-ins'. The 'buy-in' is an increasingly popular device used for altering the ownership and capital structure of companies. It became common in Britain in the 1980s. Once suitable and promising business plans have been prepared for the business in question, a management team (which may or may not consist of former managers) buys a small equity stake in the business, and this is matched with a large investment made by one or more banks or financial houses. Once the business is acquired, further finance is usually needed to put it on its feet, and this is also provided by banks or financial institutions.

The first business to be rescued in the group was the former Leyland DAF truck plant at Leyland, which was the subject of a buy-in led by local management. A rival bid for this plant had been submitted by a team led by Dan Wright, a Scot, but the locally-led bid prevailed. Wright, an engineering graduate of Glasgow University, had at one stage in his career been product development manager at the Bathgate plant, had also worked with Ford and had moved to Fleming Thermodynamics, a subsidiary of Motherwell Bridge increasingly involved in consultancy. Wright's plan was to retain the Leyland operation as an integrated chassis and components complex, but the preferred plan was for the Leyland truck plant to be sold off as a free standing assembler of vehicles.

Wright, originally called in through Fleming Thermodynamics by the Glasgow Development Agency, next turned his attention to the Albion plant and a component plant at Leyland that had not yet been sold off, Spurrier. Wright was convinced that the combined business would work if only a financial package could be arranged, but meantime it faced cash difficulties as 1993 wore on. An interim arrangement was put in place. The receiver, accountants Arthur Andersen, in an unusual move, took an equity stake on behalf of creditors in the new company taking over the Albion and Spurrier plants. In addition £4 million of regional selective assistance was provided by the Scottish Office, and £6 million was loaned by the Bank of Scotland. Thus, the new business was started. The Spurrier operation produced engine parts and chassis components, while Albion of course produced axles. The combined workforce at both plants at the time of the rescue was about 510 employees, 330 at Albion and 180 at Leyland. The employees agreed to a five per cent wage cut and the introduction of flexible work practices.

The principal customer of the new business was the Leyland and DAF truck plants, but it was clear at an early stage that the customer base had to be widened to use up capacity and for long term financial viability. How was this to be done? The new two-plant business had no product development staff, no marketing department, and its financial functions were geared to its previous role as part of an integrated group. Appointments were therefore made to these departments, and new products soon began to be designed. The circumstances were propitious: the European economy was beginning to pick up, and it was becoming more acceptable than ever before for manufacturers of vehicles to outsource components and sub-assemblies than at any time in the past. The industry had come to accept what amounted to a form of self-imposed rationalisation. The new business required a name and an identity. There was little doubt that the Albion name still carried weight and that it stood for durability and quality. The operation was to be known as Albion Automotive, and was later re-registered at Companies House as Albion Auto Industries. Thus, Albion was resuscitated. The distinctive scroll sign that had served the original Albion company for most of its existence was adopted as a logotype.

Another piece of the jigsaw remained. The new Albion board initially consisted of Dan Wright and Jim Hastie, former operations director for components at Leyland DAF and sometime finance manager at Bathgate, together with Stephen Spencer, formerly of G.E.C. It felt that it would like to acquire the crankshaft machining facility that stood in its Leyland plant but which was owned by a leasing company. Negotiations began for its acquisition. This business supplied crankshafts to DAF, and employed 30 workers. The underlying rationale on which the receivers had been working was that the newly reconstituted businesses taking over parts of the Leyland DAF empire should be launched with the security of inter-plant component supply contracts as a basis from which to move forward. The acquisition of crankshafts was in line with this rationale, and in due course took place.

Meantime, work was in hand to produce a new range of products for the Albion plant. The new product development department, with the aid of its advanced CADCAM systems, worked towards September 1995 as a launch date for a new family of light-to-medium front and rear axles, to be known as 'Truckdrive'. Leyland Trucks had already committed itself to taking this axle for its new DAF 55 truck. The new model of axle would be designed on the hypoid system of gear configuration rather than on the spiral bevel principle which characterised earlier axles made at Albion. The change was made to reduce noise, parts employed and weight. In due course a range of sizes was designed, suited to different loadings and wheel dimensions, and encompassed front and rear axles as well as four-wheel drive variants. The range was successfully launched on time in Glasgow before an audience representing Europe's best known truck producers.

By late 1995, Albion had also acquired Farington plant near Leyland, previously purchased by Volvo from Leyland, which made transmission products. This together with sales growth had taken the combined workforce up to 1,160 people. The acquisition brought with it an added bonus: an order for power take offs and driveline flanges from Volvo, to whom Albion was to become the main supplier of these components. Albion's in-house newsletter 'Componentry in Motion', described the acquisition as 'the perfect match', Farington's expertise in gears, P.T.O.'s gearboxes and like products dovetailing with Albion's axle output. Farington possessed 40 high-technology C.N.C. machines as well as some 400 conventional machine tools.

Earlier in 1995, an order for £25 million was taken from Perkins engines, for crankshafts. Some 35,000 axles per year were now being built at Scotstoun, and well known names such as Seddon Atkinson and J.C.B. were buying components from the new company. By March 1995, the company had planned on a loss of £1.5 million, but thanks to a better-than-expected truck market, the company had reached breakeven, three years ahead of plan. In June 1995, the receiver sold his equity stake for £1 million, 18 months ahead of schedule. At this stage Candover, the private equity investor put in about £19 million to refinance the debt and finance future development. By the end of 1995, Albion was anticipating annual sales of about £70 million.

By September 1996, Dan Wright had left Albion, and Jim Hastie was appointed chief executive in his place. By this time, a remarkable turnround had been effected. A textbook receivership had been accomplished thanks to the gargantuan efforts of all concerned. The company had remained profitable in the year to March 1996, although only marginally so.

What had happened on the shop floor? At the start of Albion's reincarnation, the South Works contained 880 machine tools ranging from state-of-the-art computerised workstations to 50 year-old mechanical relics. These were systematically weeded out. Production lines were reorganised for the recently introduced product ranges, and the floor, freshly painted, is today as clean as it has ever been. Links with the past are still there. Small numbers of 'indestructible' hub reduction axles, designed by Albion in the early 1960s, are still produced for the third world replacement market, although these may eventually be sub-contracted.

Many of the workers and some of the staff have been in Albion since the 1960s and feel more secure today than they have done for some time, and while the 'reverse takeover' of recent years means there are more Albion employees south of the border (800) than north of it (350), at Scotstoun, at least, the Albion tradition lives on.

CHAPTER TWELVE

Inextinguishable Rays: Albion and the Biggar Connection

Whatever may befall Albion in the future, it will never be forgotten. The Royal Scottish Museum in Edinburgh possesses one of the first Albions ever built, and the Museum of Transport in Glasgow has several vehicles in its collection. The largest repository of Albion memorabilia is, however, to be found in Biggar in the Albion Archives, presently housed in the former police station premises in Edinburgh Road, at the east end of the town. Here may be found a collection of vehicle radiator badges, medals awarded to Albions in the early vehicle trials of the Edwardian era, oil portraits of the founder-directors, a mounted Albion/Murray patent lubricator, models of Albion vehicles and many other tangible mementos of the company. It is here, too, that a large documentary collection of Albion's business records, on which this book has depended, may be consulted. Its contents are listed in the bibliography.

It is entirely appropriate that these irreplaceable effects are held in Biggar. Its people were from the very first conscious that one of their youth, Thomas Blackwood Murray, was a rising man. News of his graduation from Edinburgh University and his engineering achievements during the 1890s could hardly be kept secret in such a small and closely knit community even if he had wanted this. He did not, and whilst in Biggar, he played a full part in the life of the town. It was here that on St. John's day in 1899 he was installed as Master of the local Masonic lodge, whose history extended back to 1727. Just one month later he married Hettie Rusack, daughter of the owners of the new Rusack's Marine Hotel in St. Andrews, and it was in Hettie's home town that the ceremony took place. Biggar thenceforth took an interest in the new Mrs. Blackwood Murray, who would thereafter be a frequent visitor. Murray had himself worshipped at Biggar parish church, one of the last gothic churches to be built in Scotland after the reformation. He had been a Sunday school teacher, had been in the church choir, and his sister played the church organ as well as the pipe organ at Heavyside that was powered by the mill dam.

Lest it be thought that N. O. Fulton is by implication neglected by emphasizing the centrality of Biggar, his connections with the town must also be pointed out. He was such a frequent visitor at the turn of the century that he made the acquaintance of his future wife, a sister of Murray, there, and it was to Biggar that the two men came when road testing the first Albion car in 1900. On one such run to Lyne Mill the two men had taken advantage of Fulton's U.S. experience by fitting spoked wheels with American pneumatic tyres, but one of

12.1 T. B. Murray and his new wife in Paris in 1900.

these punctured and they had to stay the night to fix it. The wheels subsequently collapsed at the bridge at Spittal, and the result was the fitting of artillery type wooden wheels with solid tyres to the first Albions.

Jobs at Albion were sometimes available to locals. The first factory at Finnieston Street was fitted out by Biggar masons, plumbers and gasfitters, and Alexander Donaldson, the company's purchasing manager and subsequently a director, was from nearby Carnwath. He had studied engineering in Glasgow, and met Murray in a train. Murray gave him his card and he started work and went on to make his career at Albion.

Occasionally there was gossip when, for example, in later years the Blackwood Murrays would be in residence at Heavyside after the death of John Lamb Murray. Concern was expressed when Hettie was away and Murray's secretary, Mrs. Cameron, came down to Biggar, or when Murray was alone in the house with the cook, a young lady from Biggar who 'lived in' at Heavyside. There is no incriminatory evidence, but the gossip perhaps suggests that Biggar residents, rightly or wrongly, did not regard Murray as being incapable of indiscretion. Such suspicions would never have been entertained in the case of Fulton.

It must have been with some pride, too, that a convoy of A10s destined for France in 1915 was received in Biggar, and the final symbolic bonding of Albion,

THOMAS BLACKWOOD MURRAY

SCOTTISH MOTOR PIONEER

BORN AT BIGGAR
22ⁿᵈ APRIL
1871

The first ALBION car 1900

CENTENARY EXHIBITION

Biggar Museum Trust

INVITES

to the Opening of the Exhibition
by

THE LORD CLYDESMUIR

(Chairman, Scottish Council Development &
Industry)

at Central Garage, Biggar,
on Monday, 6th September, 1971
at 3 p.m.

R.S.V.P. to Brian Lambie, Gladstone Court
Museum, Biggar, Lanarkshire.

12.2 Invitation to the T. B. Murray centenary exhibition, Biggar, 1971.

Biggar and Murray of course occurred at the latter's burial, when a brand new Albion lorry served as his hearse.

It was in 1968 that Biggar received its first museum, Gladstone Court, which was designed to recreate the atmosphere of a street of shops as it might have been in the earlier decades of the century. As with subsequent ventures in a similar vein, this came about through the initiative of Brian Lambie, proprietor of a Biggar ironmongery business and town provost. In 1971, he was the prime mover in arranging an exhibition to mark the centenary of T. B. Murray's birth at Heavyside. It ran for two weeks in a borrowed car showroom.

It was opened by Lord Clydesmuir and was attended by Murray and Fulton family representatives from the U.K. and South Africa as well as senior Albion management. A tape recorded message from Hettie Blackwood Murray, then 95, gave a rousing welcome to all present. Included in the display was a newly refurbished dog cart and an A6 pleasure car, loaned by the Scotstoun works. A concourse of veteran vehicles followed on the final day. This raised the possibility of an annual event, and after a false start in 1972, the Biggar Rally was launched with about 20 vehicles in 1973. As a result of the hard work of Biggar stalwart Meg Sykes, it has now grown to include some 300 vehicles plus stationary engines and agricultural machinery. The strong commercial vehicle contingent includes many Albions from Britain and Ireland.

193

12.3 J. B. Murray, R. C. Dougal and L. Capaldi at the Murray Centenary exhibition, 1971.

Brian Lambie and the newly formed Biggar Museum Trust continued their heritage work throughout the 1970s: next, the Biggar Gasworks, opened in 1839, was saved and responsibility for its preservation taken over by Historic Buildings and Monuments and the National Museums of Scotland. By 1981, a derelict covenanter's house had been lifted stone by stone from its location at nearby Wiston and rebuilt as a museum in the heart of Biggar. In 1983, the old Free Kirk at Broughton, where John Buchan's father was minister, was converted into a John Buchan museum. Prior to this, the ongoing dialogue between the Biggar Museum Trust and the Albion plant at Scotstoun had resulted in a set of Albion memorabilia building up, and plant director Jim McGroarty completed the collection by donating the board minutes and the residue of the business records. These were at first kept in a further acquisition, Moat Park Church, which was converted into a heritage centre opened by the Princess Royal in 1988, but were subsequently moved into an adapted sausage factory prior to being placed in the former police station. A dedicated Albion museum, where the trust can keep its recently acquired 1902 Albion dog cart, archives and other exhibits, is planned, but depends on the availability of funding. The archive in its previous and present locations has served a very important purpose for Albion owners. It has a collection of vehicle handbooks and a huge volume of photographs and glass negatives of early vehicles and parts which were donated by the Scotstoun plant. It also acts as a kind of mecca for old employees and their descendants, who traverse the world to visit.

12.4 The Scotstoun works date stone re-installed at Biggar, 1989, with, left to right, Brian Lambie, Lady Cowie (granddaughter of T. B. Murray), Grizel Hoyle and Jim McGroarty.

While all this dramatic heritage activity was afoot, the first issue of the Albion (Owners) Club magazine was published in 1985 and sent to 150 potential members. This was to be produced quarterly and was aimed at an obviously growing community of Albion owners and devotees. As a result of the Biggar Rally and personal recommendation, the number to whom it is circulated now exceeds 400, of whom about 30 are in Australia. There are members in Canada, South Africa and Europe. Through the magazine, devotees can read reports of the restoration of Albion vehicles, which are often transformed from heaps of scrap into gleaming, pristine condition. A 'notes and queries' section gives brief details of Albion spares sought or offered by owners, progress at Biggar on Albion matters, accessions to the collection, details of people and so on. Articles on the restoration of Albions abroad also appear, copiously illustrated as in the case of U.K. restorations. One such report featured the restoration of an Albion 'Venturer' bus in Australia.

For those more interested in the social side of Albion affairs, a good picture of life at Albion emerged in a number of recent articles by ex-employee Bob Coutts, whose experience as an apprentice and subsequently as a fitter at Scotstoun and in Albion depots was recalled in remarkably vivid detail. Similar accounts by Jack Butterworth, John B. Clark and Jim Higgins have also appeared, as did the reminiscences of Malcolm Beaton, head draughtsman at Scotstoun until 1964. This material was interspersed with reprints from the commercial press of road tests of Albions from the 1920s to the 1950s, and with excerpts from the potted history of Albion taken from the *Radiator* and from the company's historical albums, now in the archive. As well as the above material, reviews of books on the motor industry, reports on scale models of Albions that can be purchased and on the Biggar Rally have been included, inter alia. A recent edition gave details of a planned centenary rally, to be held in 1999. As with many other initiatives connected with Albion and Biggar, the magazine is sustained by arch-enthusiast Brian Lambie (now an M.B.E.) aided and abetted by a willing team of helpers. The magazine sums up the deep rooted affection that exists throughout the world for an institution that approaches its century, for its products and for the experiences that both have given over the period.

It has been stated, incontrovertibly, in this writer's opinion, that the most important organisational form in today's society is the business corporation. So far as Scotland is concerned, Albion, by reason of its economic contribution, the usefulness of its products and its longevity, must surely be regarded as one of the most significant. In the circumstances, it is gratifying that so many components of the Albion heritage are in such good and caring hands.

Bibliography

Unpublished Sources, Albion Archives, Biggar

Albion Motor Car Company Ltd., Register of Stockholders, 1902–14.

Albion Motor Car Co. of Australasia Ltd., Minute Book, 1921–39

Albion Motor Car Company of Australasia, Register of Members, 1921–28.

Albion Motor Car Co. of Australasia Ltd., Register of Seals.

Albion Overseas Ltd. Register of Members and Secretaries, 1948–52.

Albion Register of Directors, 1902–14.

Albion Register of Directors, 1915–46.

Albion Register of Shareholders, 1946–77.

Canteen and Restaurant Cash Book, 1917.

Design Sketch Book, 1900–21.

Foreign Order Book, 1920–27.

General Minute Book of Directors Meetings 1a, 1925–27.

General Minute Book of Directors Meetings 1b, 1924–40.

General Minute Book of Directors Meetings 2, 1909–51.

General Minute Book of Directors Meetings 3, 1914–18.

General Minute Book of Directors Meetings 4, 1918–20.

General Minute Book of Directors Meetings 5, 1920–22.

General Minute Book of Directors Meetings 6, 1923–26.

General Minute Book of Directors Meetings 7, 1926–28.

General Minute Book of Directors Meetings 8, 1928–29.

General Minute Book of Directors Meetings 9, 1929–31.

General Minute Book of Directors Meetings 10, 1931–33.

General Minute Book of Directors Meetings 11, 1933–36.

General Minute Book of Directors Meetings 12, 1936–38.

General Minute Book of Directors Meetings 13, 1938–41.

General Minute Book of Directors Meetings 14, 1941–45.

General Minute Book of Directors Meetings 15, 1945–47.

General Minute Book of Directors Meetings 16, 1947–49.

General Minute Book of Directors Meetings 17, 1949–51.

General Minute Book of Directors Meetings 18, 1951–53.

General Minute Book of Directors Meetings 19, 1953–55.

General Minute Book of Directors Meetings, 1955–57.

General Minute Book of Directors Meetings, 1958–58.

List of Drawings, 1901–09.

Memorandum and Articles of Association, Albion Motor Car Company Ltd., 1902.

Minute Book of Albion Overseas Ltd., 1970–74.

Minutes of Managers Meetings, 1915–21.

Miscellaneous Deliveries Book, 1912–15.

Production notebook, 1912–20.

'Radiator' — Albion Works' Magazine, Mock-up, 1922.

'Radiator' — Albion Works' Magazine, 1928–37.

Report on Visit to U.S.A. Motor Works, W. P. Kirkwood, R. Ayton, March/April 1949.

Report and Accounts, 1914–50.

'Sunrise', Albion Trade Union Magazine, 1945–47.

Typescript Board Minutes, 1970–75.

Typescript Memoir 'Having Said All That' by H. W. Fulton, 1964.

Vehicle Type and Chassis No. List, 1901–69.

Other Unpublished Sources

Companies House, Edinburgh, Albion Report and Accounts, 1951–70.

Journals, Periodicals and Newspapers

Albion Automotive Product Brochures, 1994–96.

Albion Owners' Club *Newsletter* Nos. 1–42, 1985–97.

Biggar Veteran and Vintage Rally Programme, 1996.

Classic Trucks Focus '100 Years of Leyland', Ian Allan, Littlehampton, Summer 1996.

Commemorative Programme — 100 Years of Truckbuilding in Leyland, 1996.

Componentry in Motion, Albion Automotive Newsletter, Nos. 1–5.

Economic Review 'The Motor Industry', pp 32–4, 4 March 1987.

Financial Times, 12 December 1995, 'Back from the Brink'.

French, M., 'Manufacturing and Models in the Commercial Vehicle Industry: the Albion Motor Company, 1920–56', *The Journal of Transport History*, 1994, Vol. 15, pp 59–77.

Herald, 9 October 1993, 'Albion Workers Face Pay Cut to Create New Company'.

Herald, 2 July 1994, 'A Name that Still Spells Quality'.

Herald, 14 March 1995, 'Heavy Order Book Sees Albion Back in Profit'.

Herald, 5 September 1996, 'Wright Set to Depart Albion'.

MacDonald, A. C. and Browning, A. S. E., 'History of the Motor Industry in Scotland', *Proceedings of the Institute of Mechanical Engineers (Automobile Division)*, 9, 1961, pp 319–37.

McKinstry, S., 'The Albion Motor Car Company: Growth and Specialisation 1899–1918', *Scottish Economic and Social History*, Vol. 11, 1991, pp 36–51.

McKinstry, S., 'Argyll Motors Ltd.: A Corporate Post-Mortem', *Business History*, Vol. 37, No. 4 (1995), pp 64–84.

McKinstry, S., 'Financial management in the Early Scottish Motor Industry', *Accounting, Business and Financial History*, Vol. 3, No. 3, 1993, pp 275–90.

1974 Report and Accounts, British Leyland.

Saul, S. B., 'The Motor Industry in Britain to 1914', *Business History*, V: 23:44.

Times, 27 November 1993, 'Albion Axle Deal Saves 500 Leyland DAF Jobs'.
and:
Motor Traction, Motor World, Commercial Motor, passim.

Books

Baldwin, N., *The Illustrated History of Albion Vehicles*, Sparkford, 1988.

Booth, G., *The British Motor Bus: An Illustrated History*, Littlehampton, 1986.

Checkland, S. and Slaven, A., (Eds.), *Dictionary of Scottish Business Biography*, Vol I, Aberdeen, 1988.

Edwardes, M., *Back from the Brink: an Apocalyptic Experience*, London, 1983.

Hood, N. and Young, S., *Multinationals in Retreat: the Scottish Experience*, Edinburgh, 1982.

Hume, J. and Moss, M., *Beardmore: the History of a Scottish Industrial Giant*, London, 1979.

Lewchuk, W., *American Technology and the British Vehicle Industry*, Cambridge, 1987.

Maxcy, G. and Silberston, A., *The Motor Industry*, London, 1959.

Payne, P. L., *Studies in Scottish Business History*, Glasgow, 1967.

Pollard, S., *The Development of the British Economy*, (3rd Edition), 1914–80, London, 1989.

Sked, A. and Cook, C., *Post-War Britain: a Political History*, (4th Edition), London, 1993.

Slaven, A., *The Development of the West of Scotland, 1750–1960*, Glasgow, 1975.

Thoms, D. and Donnelly, T., *The Motor Car Industry in Coventry Since the 1890s*, London, 1985.

Turner, G., *The Leyland Papers*, London, 1971.

Wilks, S., *Industrial Policy and the Motor Industry*, Manchester, 1984.

Wood, J., *Wheels of Misfortune: the Rise and Fall of the British Motor Industry*, London, 1988.

An Albion Chronology

1898 N. O. Fulton goes to U.S.A. to gain motor industry experience.

1899 Albion Motor Car Company established at Finnieston, Glasgow by T. B. Murray and N. O. Fulton, as partnership.

1901 Silver Medal awarded to 8 h.p. Albion of Walter Creber in Automobile Club 500 Miles Trial run by the Glasgow International Exhibition.

1902 J. F. Henderson BSc joins Albion and it is converted to a private limited company with T. B. Murray as chairman.

1903 Works move from Finnieston to Scotstoun.
12 h.p. Albion awarded Silver Medal in Automobile Club Reliability Trials.

1904 Lacre Motor Car Company of London awarded sole concession for sale of Albion products in England and Wales.

1905 Gold Medal awarded to Albion in Scottish Reliability Trials, 1905.

1906 24 h.p. car, Albion's only chassis designed for pleasure use alone, is launched.

1907 H. E. Fulton joins Albion as a director.
Showroom opened at 88 Mitchell Street, Glasgow.
Gold Medal awarded to two Albions in the Industrial Vehicle Trials held by the Royal Automobile Club.
Silver Medal awarded to 24 h.p. Albion Car in Vapour Emission Trials.

1908 Albion wins Scottish Cup for best petrol consumption in Scottish Reliability Trials.
Dispute with Lacre referred to arbitrator.

1909 Agreement with Lacre terminated.

1910 London depot opened.
Night shift started at Scotstoun.

1911 Manufacture of 24 h.p. car terminated.

1912 Manchester Depot opened.

1913 Glasgow showroom closed and office opened instead.
Last pleasure car manufactured.
Construction of first phase of ferro-concrete block.

1914 Branch office opened at Liverpool.
Outbreak of war. War Department 'impress' all Albion vehicles in workshops and from Albion users throughout the country. Plant given over to production (mainly of 32 h.p., 3 ton wagons) for the war effort.
Second phase of ferro-concrete block constructed.

1915 Albion becomes a public company.

1916 Dilution of labour introduced.
Shell manufacture begins.

1917 DSc degree awarded to T. B. Murray.

1918 Albion/Murray lubricator fitted to Dragonfly aircraft.
Armistice signed with Germany on 11 November.
Debenture issue of £200,000.

1919 Birmingham office opened.
Used Albions purchased from War Department for reconditioning and sale.

1920 Trade recession begins.

1921 Recession continues.
Miners' strike from 1 April to 29 June.
Australian subsidiary established.
Manchester repair shop opened.
London repair shop opened at Willesden.

1922 Recession continues.
Market flooded with ex-military vehicles.
Birmingham service and repair depot opened.

1923 Recession begins to lift.
'Viking' charabanc launched at Olympia Motor Show.

1924 Branch office opened at Leeds.
Semi-jubilee celebrated with concessionaires in Glasgow.

1925 'Overtype' lorry launched.
Bristol office opened.

1926 General Strike from 3 to 13 May.

1927 The Prince of Wales visits Albion.
Albion produces first six-wheeler.
Yoker recreation ground developed.

1928 East building extension to works begun.

1929 Death of T. B. Murray at Monthey, Switzerland on 11 June. N. O. Fulton
succeeds him as chairman.

1930 Depression begins.
King orders 30 cwt. Albion for Balmoral.

1931 Depression continues.
Factory reorganisation takes place.
Name of company altered to Albion Motors Ltd.

1932 Depression continues.
Salter Report increases taxes on transport vehicles.
First Albions fitted with diesels delivered.
Edinburgh sales and service depot opened.
Debenture issue fully repaid.

1933 Depression lifts.
Salter Report causes radical redesign of chassis to take advantage of tax and
speed regulations.
Liverpool office opened.

1934 Royal Warrant granted to Albion, which had by this point supplied three vehicles to H.R.H.

1935 N. O. Fulton dies on 27 July at Milngavie.
Halley's premises at Yoker acquired.
Westwards extension of works begins.

1936 H. E. Fulton dies, at Bearsden.
Norwich office opened.
Lincoln office opened.

1937 Service depot in Johannesburg begun.
New heat treatment facility started.
Albion Motors Overseas Ltd. was started to coordinate Albion business in Australia, New Zealand and South Africa.

1938 Hull office opened.
Works employees granted holiday pay under new scheme.

1939 Air raid shelters constructed at Scotstoun.
War declared on 3 September.
Production of largely heavy models for the War Department begins.

1940 Birmingham, Willesden and Manchester premises suffer air raid damage.
Dilution of labour begins.
Pistol manufacture begins at Yoker.

1941 J. F. Henderson dies on 30 May at Shandon.
Experimental department destroyed in air raid with minor damage caused to main works glazing.

1942 Torpedo engine manufacture begins.
Pistol and sten gun barrel production in full flow.

1943 8 wheeled CX33 experimental tank tractor produced, together with 6 wheel experimental gun tractor.

1944 Field Marshall Montgomery visits Albion on 21 April.

1945 Victory in Europe declared on 8 May.
Experimental department rebuilt.

1946 Road transport nationalised by new Labour government.

1947 Production hampered by supply shortages, exacerbated by freak blizzards and floods in February and March.
Works hours reduced to 44 per week, staff 37.5

1948 Record export sales, which have priority for steel supplies.

1949 Scottish Motor Exhibition revived.
50 per cent of output exported.

1950 Introduction of purchase tax on vehicles and increases in fuel tax slow down the market.

1951 Merger with Leyland Motors Ltd. completed on 31 July.
Belfast office closed down.

1952 Albion Cuthbertson Water Buffalo invented. Albion to produce engines.

1953 Rationalisation of models within Leyland group begins.

H. W. Fulton, son of N. O. Fulton and managing director since 1945, becomes deputy managing director of Leyland Motors.

1954 Service department transferred to Yoker premises.
      Leyland acquires Scammell.

1955 Legislation permits four wheelers to operate at a G.V.W. of six tons and an overall width of eight feet.
      Chassis designs altered to comply.

1956 Credit restrictions by government affect business.
      Suez crisis causes fuel shortage, affecting orders.

1957 New orders fall in wake of Suez crisis.
      H. W. Fulton resigns and is succeeded by Stanley Markland as managing director.

1958 Trade depression, exacerbated by anti-inflation measures and difficulties abroad, depresses the industry.
      Sir Henry Spurrier becomes chairman of Albion.
      New, rationalised Chieftain range introduced with spiral bevel rear axle and hub reduction gears, followed later in the year by a rationalised Clydesdale.

1959 Depression lifts.
      Conservative government wins general election and lifts fears of renationalisation of transport, which it had earlier denationalised.

1960 Record sales. Property acquired for expansion at Gooseholm Farm, Dumbarton (never used).

1961 New gear factory completed at corner of South St. and Balmoral St.
      Launch of Lowlander chassis marks Albion's re-entry to double deck bus market.
      Albion cumulative vehicle production exceeds 100,000 (103,396).

1962 Market depressed for part of year.
      Leyland acquires A.E.C.

1963 D. G. Stokes succeeds Spurrier as chairman of Albion and Stanley Markland resigns.

1964 Ergomatic cab introduced.
      Recessionary trend lifts.

1965 Rear-engined Viking single deck bus chassis introduced.

1966 New moving assembly track, part of a £2 million extension, opened by Princess Margaret on 15 February.

1967 Sales reach all time high with help of new facilities.

1968 Leyland Motors takes over B.M.H. to form British Leyland.

1969 As part of a £2¼ million expansion programme, Albion takes over former Harland and Wolff premises across South Street, known henceforward as South Works.

1970 Albion loses its limited liability status and becomes part of British Leyland Truck and Bus Division.
      Chassis cumulative output exceeds 150,000 (157,446).

1971 New axle line begins in South Works.
      Albion log book (historical record) ceases.

1972 Albion vehicles now utilise Bathgate produced cab, are badged with Leyland name, but can still be identified by model names, e.g. Chieftain, Clydesdale, Reiver, Viking.

1980 Chassis production transferred to Bathgate.
Albion becomes components factory.

1987 Becomes part of DAF of Netherlands.

1988 Albion North Works demolished.

1993 DAF collapses, leaving its British operations in receivership.
Albion Automotive commences as independent company.

APPENDIX TWO
Financial Details

Albion Summarised Accounts

(*Source:* Companies House, Albion Archives)

£000s	1914	1915	1916	1917	1918	1919	1920
Fixed assets	97	134	134	139	130	155	302
Investments	-	69	169	62	29	203	175
Stock	137	188	238	334	383	493	671
Debtors	43	67	98	182	219	127	152
Cash	32	42	15	9	39	72	4
	309	500	654	726	800	1050	1304
Loans & Creditors	36	157	288	316	329	291	535
Debentures					18	189	179
Issued Capital	212	262	263	263	263	395	502
Reserves	61	81	103	147	190	175	88
	309	500	654	726	800	1050	1304
Profit (Loss) After Tax	45	59	50	70	76	97	63
Dividend	16	28	34	34	34	45	8

	1921	1922	1923	1924	1925	1926	1927
Fixed Assets	373	357	359	335	336	348	364
Investments	15	12	11	10	43	44	18
Stock	391	284	232	303	371	412	469
Debtors	106	68	81	100	121	157	185
Cash	3	14	71	73	39	22	29
	888	735	754	821	910	983	1065
Loans & Creditors	201	64	76	105	154	178	197
Debentures	168	154	135	121	90	88	85
Issued Capital	502	502	502	502	502	502	502
Reserves	17	15	41	93	164	215	281
	888	735	754	821	910	983	1065
Profit (Loss) After Tax	(44)	6	45	89	118	106	122
Dividend	8	8	26	36	45	45	58

£000s	1928	1929	1930	1931	1932	1933	1934
Fixed assets	429	426	385	369	376	358	350
Investments	17	39	46	103	110	142	97
Stock	513	593	432	348	322	286	379
Debtors	178	115	126	130	102	154	121
Cash	29	24	55	30	28	37	87
	1166	1197	1044	980	938	977	1034
Loans & Creditors	278	226	104	117	113	169	187
Debentures	69	52	35	16	-	-	-
Issued Capital	502	548	548	548	548	548	548
Reserves	317	371	357	299	277	260	299
	1166	1197	1044	980	938	977	1034
Profit (Loss) After Tax	94	72	50	(16)	(10)	(6)	46
Dividend	58	60	39	8	8	8	29

	1935	1936	1937	1938	1939	1940	1941
Fixed Assets	368	373	380	406	423	436	431
Investments	84	105	107	124	292	208	193
Stock	560	486	708	546	669	1042	1246
Debtors	147	162	216	219	203	215	207
Cash	28	283	146	304	120	102	64
	1187	1409	1557	1599	1707	2003	2141
Loans & Creditors	268	271	361	316	397	765	729
Debentures	-	-	-	-	-		
Issued Capital	548	631	631	631	631	631	631
Reserves	371	507	565	652	679	607	781
	1187	1409	1557	1599	1707	2003	2141
Profit (Loss) After Tax	83	123	148	159	111	125	122
Dividend	49	83	70	83	70	70	70

	1942	1943	1944	1945	1946	1947	1948
Fixed assets	429	396	357	338	360	380	559
Investments	152	135	132	151	373	549	-
Stock	1456	1537	1482	1510	1315	1618	2526
Debtors	137	128	78	160	406	300	392
Cash	89	264	141	147	102	110	35
	2263	2460	2190	2306	2556	2957	3512
Loans & Creditors	796	927	605	655	723	958	960
Debentures	-	-	-	-	-	-	-
Issued Capital	631	631	631	631	631	631	731
Reserves	836	902	873	1020	1202	1368	1821
	2263	2460	2109	2306	2556	2957	3512
Profit (Loss) After Tax	127	128	123	139	213	250	246
Dividend	69	70	70	70	82	82	54

£000s	1949	1950	1951	1952	1953	1954	1955
Fixed Assets	582	635	591	705	818	764	874
Investments	-	-	-	-	-	86	86
Stock	2289	2463	2729	3344	2972	2881	3024
Debtors	536	596	832	968	1132	1246	1316
Cash	134	245	182	115	83	101	190
	3541	3939	4334	5132	5005	5078	5490
Loans & Creditors	872	1025	1224	1741	1843	1672	1775
Debentures	-	-	-				
Issued Capital	731	731	731	731	731	731	731
Reserves	1938	2183	2379	2660	2431	2675	2984
	3541	3939	4334	5132	5005	5078	5490
Profit (Loss) After Tax	185	228	221	308	235	304	306
Dividend	54	51	33	51	70	70	87
Turnover					7574	6382	7296

£000s	1956	1957	1958	1959	1960	1961	1962
Fixed assets	983	1089	1081	1043	1077	1356	1581
Investments	86	78	81	80	7	7	7
Stock	4276	3624	3969	4215	4124	5123	4930
Debtors	1271	1619	1732	1714	2444	2492	2629
Cash	25	291	61	51	329	126	5
	6641	6701	6924	7103	7981	9104	9152
Loans & Creditors	2676	2788	3089	3141	3497	4072	3862
Debentures			-	-	-	-	-
Issued Capital	731	731	731	731	731	731	731
Reserves	3234	3182	3104	3231	3753	4301	4559
	6641	6701	6924	7103	7981	9104	9152
Profit (Loss) After Tax	306	230	(96)	230	647	658	424
Dividend	108	(e)108	34	114	133	133	180
Turnover	7914	n.a.	7026	8972	11562	12180	11911

£000s	1963	1964	1965	1966	1967	1968	1969
Fixed Assets	1394	1440	1726	1873	1863	1525	1752
Investments	7	7	7	7	7	7	7
Stock	3996	4740	6593	7127	7109	7185	7632
Debtors	7478	6918	7252	6646	6811	8025	8282
Cash	29	78	43	4	4	51	206
	12904	13183	15621	15657	15794	16793	17879
Loans & Creditors	7319	7357	9030	7811	8681	11196	11782
Debentures	-	-	-	-	-		
Issued Capital	731	731	731	731	731	731	731
Reserves	4854	5095	5860	7115	6382	4866	5366
	12904	13183	15621	15657	15794	16793	17879
Profit (Loss) After Tax	412	829	1171	704	668	1401	1304
Dividend	137	640	465	n.a.	528	438	800
Turnover	13361	15423	15197	16025	17321	20297	20687

Ratio Analysis

	Liquidity (Current Assets/ Current Liabilities)	Profitability (Profit after tax as % of Capital & Reserves)	Re-investment (Profit after tax retained %)	
1914	5.9/1	16.5	64	
1915	2.3/1	17.2	53	
1916	1.8/1	13.7	32	
1917	1.9/1	17.1	51	
1918	2.0/1	16.8	55	
1919	3.1/1	20.6	54	
1920	1.9/1	10.7	87	
1921	2.6/1	(8.5)	-	
1922	5.9/1	1.0	-	
1923	5.2/1	8.3	42	
1924	4.6/1	15.0	60	
1925	3.7/1	17.7	62	
1926	3.6/1	14.8	58	
1927	3.6/1	15.6	52	
1928	2.7/1	11.5	38	
1929	3.2/1	7.8	17	
1930	5.9/1	5.5	22	
1931	4.3/1	(1.9)	-	
1932	4.0/1	(1.2)	-	
1933	2.8/1	(0.7)	-	
1934	3.1/1	5.4	37	
1935	2.7/1	9.0	41	
1936	3.4/1	14.7	33	
1937	3.0/1	12.5	53	
1938	3.4/1	12.4	48	
1939	2.5/1	8.5	37	
1940	1.8/1	10.1	44	
1941	2.1/1	10.1	43	
1942	2.1/1	8.6	46	
1943	2.1/1	8.7	45	
1944	2.8/1	7.8	43	
1945	2.8/1	8.4	50	
1946	2.4/1	11.6	62	
1947	2.1/1	12.5	69	
1948	3.1/1	9.6	78	
1949	3.4/1	6.9	71	
1950	3.2/1	7.8	78	
1951	3.1/1	7.1	85	9 months
1952	2.5/1	9.1	83	
1953	2.3/1	7.4	70	
1954	2.6/1	8.9	77	
1955	2.6/1	8.2	72	

	Liquidity (Current Assets/ Current Liabilities)	Profitability (Profit after tax as % of Capital & Reserves)	Re-investment (Profit after tax retained %)
1956	2.1/1	7.7	65
1957	2.0/1	5.9	53
1958	1.9/1	(2.5)	–
1959	1.9/1	5.8	50
1960	2.3/1	14.4	79
1961	1.9/1	13.1	80
1962	2.0/1	8.0	58
1963	1.6/1	7.4	67
1964	1.6/1	14.2	23
1965	1.5/1	17.8	60
1966	1.8/1	9.0	n.a.
1967	1.6/1	9.4	21
1968	1.4/1	25.0	69
1969	1.4/1	21.4	39

N.B. Re-investment figures after 1950, when Leyland took Albion over, are less reliable, since they are bound up with group treasury arrangements from this point onwards.

Albion Output Statistics

Year	Vehicles	Engines	Gearboxes	Axles	Other
1900	1				
1901	21				
1902	35				
1903	33				
1904	65				
1905	n.a.				
1906	221				
1907	248				
1908	n.a.				
1909	n.a.				
1910	282	39			
1911	354	50			
1912	n.a.				
1913	554	91			
1914	591	95			
1915	927	84			
1916	1454	82			
1917	1689	75			23237 shells
1918	1843	85			50565 shells
1919	1703	40			
1920	1506				
1921	166				
1922	244				
1923	606				
1924	1110				
1925	1309	94			
1926	1295	99			
1927	1489	124			
1928	1658	53			
1929	1538	30			
1930	1302	10			
1931	1271	6			
1932	1362				
1933	1380				
1934	2003				
1935	2918				
1936	4051				

Year	Vehicles	Engines	Gearboxes	Axles	Other
1937	3968				
1938	3752				
1939	3102				
1940	2492				
1941	1860				684 pistols
					31550 sten gun barrels
1942	1213	91 (torpedo)			13261 pistols
					318523 sten gun barrels
1943	1456	519 (torpedo)			7477 pistols
					80644 sten gun barrels
1944	1038	621 (torpedo)			
1945	980				
1946	1496				
1947	2040				
1948	1840				
1949	2580				
1950	2881				
1951	2570	97			
1952	2223	376			
1953	2545	547	572		
1954	3458	397	1740		
1955	3565	355	3506		
1956	3861	527	3617		
1957	2520	203	4894		
1958	2256	170	3543		
1959	3442	294	5452	739	
1960	4700	365	6238	1331	
1961	4982	282	5402	1672	
1962	4212	183	3376	1279	
1963	4885	34	4881	2188	
1964	6102	81	5774	3804	
1965	6047	58	4961	3794	
1966	5736	93	3620	2804	
1967	6312	27	3260	4103	
1968	6539	-	4669	5473	
1969	6754	-	4800	5193	
1970	7463	-	5520	5177	

N.B. Engines, Gearboxes and Axles produced figures represent units produced in addition to those included in chassis.

Source: Albion Historical Albums

Albion's Order Book
to July 1904

Car No.	Date Ordered	Date Delivered	Type	Name of Purchaser
6A	4.1.00	1.2.01	Dogcart	John L. Murray, Biggar
6B	10.9.00	26.1.01	8 h.p. dogcart	Walter Graham, Bearsden
6C	12.9.00	29.12.00	Dogcart	W. A. Verel, Newlands
6D	12.10.00	9.3.01	Dogcart	John Robertson, Galston
6E	6.11.00	8.3.01	8 h.p. dogcart	Walter Creber, Barrhead
6F	25.2.01	7.3.01	Dogcart	Robert Millar, Bearsden
13A	25.10.00	30.5.01	8 h.p. dogcart	Sir T. D. Gibson Carmichael, Castle Craig
13B	27.2.01	3.6.01	Dogcart	Dr. J. L. Howie, Annan
13C	14.3.01	6.7.01	Tonneau	Charles Robertson, Pollokshaws
13D	2.3.01	28.5.01	8 h.p. dogcart	Capt. E. Hunter Blair, Maybole
13E	5.5.01	12.6.01	8 h.p. dogcart	John Adam, Shettleston
13F	22.9.02	24.9.02	Dogcart	J. McConnell, Dumfries
20A	15.5.01	5.8.01	Dogcart	John Motherwell, Airdrie
20B	16.5.01	1.8.01	Dogcart	Mitchell Bros., Glasgow
20C	2.5.01	10.9.01	Dogcart	Lewis R. Lamont, Glasgow
20E	17.5.01	20.8.01	Dogcart	William Murray, Hillhead
20F	7.6.01	6.9.01	Tonneau	Millar & Allen (for shipment)
45A	16.8.01	11.10.01	Tonneau	Hutson
45B	16.8.01	11.10.01	Dogcart	Robert Rankin, Govan
45C	11.9.01	11.10.01	Dogcart	George Raffon, Sydney, N.S.W.
45D	19.9.01	21.12.01	Tonneau	James Potter, Pollokshields
45E	25.9.01	21.11.01	Dogcart	George Spicer, London (for shipment to S.A.)
45F	27.9.01	26.10.01	Dogcart	David Todd
50A	2.10.01	16.1.02	8 h.p. dogcart	S. Briggs Berry, Accrington
50B	18.11.01	11.3.02	Tonneau	James Mitchell, Caldercruix
50C	12.12.01	12.3.02	Tonneau	Dr. A. Veitch, Edinburgh
50D	23.12.01	27.3.02	8 h.p. tonneau	William Verel, Newlands
50E	23.12.01	7.2.02	Dogcart	John M. Ross, Pollokshields
50F	8.1.02	9.4.02	Tonneau	Roderick Morrison, Partickhill
81A	8.1.02	8.5.02	Wagonette	Millar & Allen, Glasgow
81B	8.1.02	8.5.02	Wagonette	Millar & Allen, Glasgow
81C	23.1.02	17.5.02	Tonneau	J. J. Pollock, Blanefield
81D	11.2.02	31.5.02	8 h.p. tonneau	Edward Crosber, Ibrox
81E	12.2.02	31.5.02	Tonneau	Norman Glen, Dowanhill

Car No.	Date Ordered	Date Delivered	Type	Name of Purchaser
81F	18.2.02	17.5.02	8 h.p. tonneau	D. Johnstone, Glasgow
163A	28.2.02	16.7.02	Tonneau	Jas. McFarlane, Glasgow
163B	25.3.02	19.6.02	10 h.p. tonneau	W. P. MacLay, Kilmacolm
163C	3.4.02	19.6.02	10 h.p. tonneau	J. W. Torrance, Hillhead
163D	9.4.02	1.7.02	Dogcart	Thomas Dewsbury, Leeds
163E	16.4.02	12.6.02	Chassis	Millar & Allen (for shipment)
163F	12.4.02	8.7.02	Tonneau	Sir J. Murray, Rannoch
230A	19.4.02	10.10.02	8 h.p. tonneau	T. McFarlane, Airdrie
230B	1.5.02	10.9.02	Tonneau	Millar & Allen (for shipment)
230C	2.5.02	7.8.02	8 h.p. tonneau	J. Martin Newton, Gullane
230D	31.5.02	16.8.02	8 h.p. tonneau	Mrs. Anderson, Kelvinside
230E	2.6.02	18.8.02	8 h.p. tonneau	Nicol P. Brown, Kelvinside
230F	11.6.02	5.10.02	8 h.p. tonneau	D. H. Anderson, Helensburgh
270A	12.6.02	8.10.02	8 h.p. tonneau	J. W. W. Drysdale, Kelvinside
270B	16.7.02	13.11.02	6-seat wagonette	Malay Transport Syndicate
270C	16.7.02	14.11.02	6-seat chassis	Malay Transport Syndicate
270D	16.7.02	3.12.02	chassis	Malay Transport Syndicate
270E	30.7.02	5.12.02	Wagonette	Millar & Allen (for shipment to Straits Settlements)
270F	5.8.02	5.12.02	Wagonette	Millar & Allen (for shipment to Straits Settlements)
335A	16.9.02	19.12.02	Dogcart	Herbert C. Burton, Blackheath
335B	24.9.02	25.12.02	Sport dogcart	Dr. Galloway, Singapore
335C	4.10.02	12.12.02	8 h.p. chassis	Mr. Irving, Dumfries
335D	4.10.02	12.12.02	10 h.p. chassis	A. C. Penman, Dumfries
335E	4.10.02	9.3.03	10 h.p. tonneau	D. M. Wilson, Options Ltd., London
335F	19.11.02	29.12.02	10 h.p. tonneau	R. McAlpine & Sons, Glasgow
483A	19.12.02	28.3.03	10 h.p. tonneau	George A. Mitchell, Hillhead
483B	8.10.02	6.3.03	10 h.p. tonneau	Walter Mitchell, Airdrie
483C	18.10.02	20.3.03	10 h.p. tonneau	Walton
483D	17.11.02	30.4.03	10 h.p. sport dogcart	R. Montgomery Stevenson, Edinburgh
483E	19.11.02	6.3.03	8 h.p. station cart	Finnieston Engineering Co.
483F	26.11.02	20.3.03	8 h.p. dogcart	Dr. P. B. Proudfoot, Kirkcaldy
620A	26.12.02	15.5.03	8 h.p. dogcart	Lawrence Bell, Innerleithen
620B	11.2.03	16.5.03	10 h.p. chassis	A. C. Penman, Dumfries
620C	12.2.03	24.4.03	10 h.p. chassis	E. G. Appleby, Redhill
620D	24.2.03	4.6.03	10 h.p. tonneau	E. S. W. Sitwell, Dorchester
620E	25.2.03	27.5.03	10 h.p. tonneau	Nicol P. Brown, Kelvinside
620F	2.3.03	13.6.03	10 h.p. tonneau	McGregor, Port of Menteith
662A	30.12.02	30.10.03	12 h.p. tonneau	Arch. Leitch for Scott, Cape Colony
662B	27.12.02		12 h.p. wagonette	Lawrence Bell
662C	27.12.02	16.10.03	12 h.p. tonneau	L. C. Seligmann, Glasgow for John M. Ross

Car No.	Date Ordered	Date Delivered	Type	Name of Purchaser
662D	31.1.03			S. Briggs, Bury
622E	31.1.03	24.10.03	12 h.p. tonneau	W. Smith, Accrington
662F	3.2.03	24.10.03	12 h.p. tonneau	Osborne, per Millar & Allen
662G	7.2.03	11.11.03	12 h.p. tonneau	W. Cornwall
662H	21.2.03	11.11.03	12 h.p. tonneau	R. Paton, Johnstone
662I	23.3.03	7.11.03	12 h.p. tonneau	W. Graham, Bearsden
680A				
680B	25.5.03	24.9.03	10 h.p. tonneau	F. G. Crouch, Sydney, N.S.W.
680C	25.5.03	22.7.03	10 h.p. tonneau	W. Hepburn, Glasgow
680D	26.5.03	24.8.03	8 h.p. dogcart	Mr. Strother, Selkirk
680E	27.5.03	16.7.03	10 h.p. tonneau	Robert Lean, Glasgow
680F	25.7.03	30.9.03	8 h.p. dogcart	Lawrence Bell, Peebles
740A	23.2.03	16.12.03	12 h.p. brougham	John Adam, Shettleston
740B	28.2.03	23.3.04	12 h.p. tonneau	D. Johnston, Kelvinside (London Show)
740C	10.3.03	28.12.03	12 h.p. tonneau	Arthur Wingate, Dowanhill (N.O.F.)
740D	13.3.03	23.1.04	12 h.p. tonneau	T. Clayton, Blackburn
740E	27.6.03	16.12.03	16 h.p. chassis	A. C. Penman, Dumfries (for Mrs Weatherley)
740F	30.7.03	19.12.03	12 h.p. chassis	A. C. Penman, Dumfries (for C. Lyall)
740G	2.7.03	10.03	12 h.p. tonneau	W. W. Walton, Ferryside
7A	29.7.03	3.3.04	12 h.p. tonneau	T. C. Spencer, St. Boswells
7B	13.8.03	19.3.04	12 h.p. tonneau	Hope Bell, Dumfries
7C	24.8.03	1.04	12 h.p. lorry	J. Klerck, Cape Town
7D	8.9.03	3.3.04	12 h.p. tonneau	W. R. Lester, Glasgow
7E	5.9.03	11.4.04	12 h.p. tonneau	Dr. Kelly, Pollokshields
7F	12.11.03	4.04	12 h.p. tonneau	John Wilson, M.P., Dunning
8A	11.9.03	12.5.04	12 h.p. tonneau	T. J. S. Roberts, Melrose
8B	15.9.03	16.4.04	12 h.p. tonneau	Dr. Waterhouse, Glasgow
8C	17.10.03	9.4.04	12 h.p. tonneau	P. T. S. Peat, Bearsden
8D	1.10.03	4.04	12 h.p. tonneau	Jas. McKillop, Glasgow
8E	23.10.03	12.4.04	12 h.p. chassis	Jas. Duff, Blairgowrie
8F	23.10.03	4.04	12 h.p. chassis	A. C. Penman
8G	28.10.03	19.5.04	12 h.p. tonneau	R. M. Stevenson, Edinburgh
8H	31.10.03	25.4.04	12 h.p. chassis	A. C. de la Fontaine, Dorchester
8I	12.11.03	18.5.04	12 h.p. 9-seat wagonette	W. Henderson, Doune
9A	14.11.03	21.5.04	12 h.p. tonneau	Provost Ballantyne, Peebles
9B	14.11.03	9.6.04	12 h.p. tonneau	R. W. Thorn, Glasgow
9C	15.4.04	25.6.04	12 h.p. limousine	Prof. A. Barr, DSc, Glasgow
9D	24.11.03	13.6.04	12 h.p. tonneau	J. M. Newton, Gullane
9E	24.11.03	13.6.04	12 h.p. chassis	E. G. Appleby, Redhill
9F	1.12.03	17.6.04	12 h.p. tonneau	A. Mackenzie, Edinburgh
9G	8.12.03	19.6.0	12 h.p. tonneau	M. C. Fleming, Richmond

Car No.	Date Ordered	Date Delivered	Type	Name of Purchaser
9H	10.12.03	27.6.04	12 h.p. tonneau	J. J. Spencer, Glasgow
9I	14.12.03	8.7.04	12 h.p. Lonsdale	Mrs. Macfie, Corstorphine
9J	18.12.03	1.7.04	12 h.p. tonneau	J. P. Wright, Edinburgh
9K	23.12.03	5.7.04	12 h.p. tonneau	S. C. Thomson, Dundee
9L	2.2.04	8.7.04	12 h.p. tonneau	Mrs. Rankin

APPENDIX FIVE
Albion Manpower Statistics

Year	Works	Staff	Total
1900	5	2	7
1901	29	3	32
1902	43	5	48
1903	68	9	77
1904	82	15	97
1905	n.a.	n.a.	n.a.
1906	252	31	283
1907	n.a.	n.a.	n.a.
1908	n.a.	n.a.	n.a.
1909	n.a.	n.a.	n.a.
1910	418	118	536
1911	584	158	742
1912	n.a.	n.a.	n.a.
1913	744	196	940
1914	1050	220	1270
1915	1253	248	1501
1916	1525	337	1862
1917	1791	366	2157
1918	1707	400	2107
1919	1691	433	2124
1920	1150	310	1460
1921	497	193	690
1922	527	248	775
1923	620	220	840
1924	950	310	1260
1925	1099	454	1553
1926	1213	493	1706
1927	1361	493	1854
1928	1733	601	2334
1929	1254	607	1861
1930	n.a.	n.a.	n.a.
1931	1196	538	1734
1932	913	562	1475
1933	1138	531	1669
1934	1291	539	1830
1935	1694	587	2281

Year	Works	Staff	Total
1936	1851	644	2495
1937	1968	673	2641
1938	1988	689	2677
1939	1900	699	2599
1940	2074	725	2799
1941	2866	748	3614
1942	n.a.	n.a.	4353
1943	n.a.	n.a.	4017
1944	n.a.	n.a.	3563
1945	2438	819	3257
1946	2347	864	3211
1947	2269	839	3108
1948	2225	835	3060
1949	2365	854	3219
1950	2523	863	3386
1951	2494	903	3397
1952	2296	949	3245
1953	2373	944	3317
1954	2437	939	3376
1955	2537	986	3523
1956	2734	1010	3744
1957	2336	931	3267
1958	2203	894	3097
1959	2246	799	3045
1960	2258	800	3058
1961	2486	793	3277
1962	2324	756	3080
1963	2257	698	2955
1964	2410	699	3109
1965	2399	701	3100
1966	2313	721	3034
1967	2418	697	3115
1968	2521	697	3218
1969	2507	712	3219
1970	2562	730	3292
1971	2440	688	3128

Source: Albion Historical Albums

Index

Aitken, Max (Lord Beaverbrook), 129
Albion/Cuthbertson Water Buffalo, 154
Amalgamated Engineering Union, 66,
　135–136
Andrews, David, 177
Argyll Motors, 11, 16, 31, 44–45
Arrol, Sir William, 2,13
Arrol-Johnston, 16, 50

Bank of Scotland, 45
Beardmore, William, 43
Beaton, Malcolm, 101, 196
Beveridge Report, 141
Bibby, Tom, 108
Biggar Kirk, 1, 89
Birmingham Office, 56, 73, 129
BMC Bathgate Plant, 165, 172, 173
Brown, Nicol Paton, 57, 86
Browne, Claude, 20
Bryce, David, 1
Buchanan, T. B., 85

Campbell, J. E., 146, 156
Camplin, M. J., 156
Chalmers, Ian, 99
Chrysler Corporation, 166
Clarke, A. D., 56
Clegg, Jim, 173
Clyde Workers Committee, 52, 53
Clydesdale Bank, 133, 154, 183
Coats, J. and P., 2, 13
Coutts, Bob, 196
Creber, Walter, 15
Cree, Alex, 10
Cuthbertson, J. A., 154

Dale, R. 176
Dilution of Labour Scheme, 51–52, 53
Donaldson, Alexander, 56, 84, 114, 124, 146
Dougal, Ron, 171, 173, 175
Douglas Castle, 2
Dunsmnir, Archibald, 108
Dykes, 6, 56

Edwardes, Michael, 176
Ellis, Ron, 173

Enfield Company, 128
Excess Profits Duty, 62–63

Fulton, Hugh Ernest, 28, 102, 112, 115,
　119
Fulton, Hugh W., 114, 127, 128, 130, 133,
　134, 136, 145, 153, 155, 157, 159

Gallacher, Willie, 51, 52
General Strike, 85
Gibson, J., 27
Gilchrist, P., 27, 56
Govan, Alexander, 44, 45
Graham, Walter, 9
Gray, John, 135, 138

H.M. The Prince of Wales, 86, 91
Heathhall Motor Works, 40, 43
Hennebique, François, 40
Halley Company, 5, 16, 43, 119, 120
Hartwood Hospital, 1
Hastie, Jim, 188
Heavyside, 1, 13
Henderson, John Francis, 13, 14, 29, 31, 40,
　54, 86, 114, 133
Hozier Engineering Company, 11, 16, 44,
　62
Hull Office, 116

Institution of Engineers and Shipbuilders in
　Scotland, 67

Johnston, George, 2

Kahn, Albert, 40
Keachie, David, 10, 27, 39, 56, 86, 89, 100
Kemp, James A., 108, 150, 153, 158
Kesson and Campbell, 2
King, Brown and Company, 1
Kirkwood, W. P., 156, 169

Lacre Company, 20, 24–27, 30, 32, 42, 59
Lambie, B.7 193, 196
Lang, William, 90
Lawson, John, 13, 14
Leeds Depot, 116

Lincoln Depot, 122
Livingston, Hugh, 7
London Depot, 65, 91

McFarlane, W., 56, 112, 113, 136, 146, 152
McGroarty, J., 178, 194
McIntosh, Cecil, 108
Macdonald, A. C., 156, 163, 173
MacDonald, Sir John H. A., 15, 33
MacLeod, John, 100
Madelvic Company, 12, 17, 43
Manchester Depot, 56, 73, 91, 129
Markland, Stanley, 160, 161, 166, 169
Matthew, Colonel, 45
Mavor and Coulson, 2, 13
Maxwell, James, 156
Millar, J., 27, 61
Millar, Jackson, 9, 31, 61
Millar, Jackson Jnr., 128, 139, 142, 147, 150, 152, 158, 163
Millburn Motors, 137
Milloy, J., 171, 173
Mo-Car Syndicate, 2, 11, 16, 100
Montgomery, General, 136
Murray, John Lamb, 1, 3, 7, 13, 192

Napier, J. S., 43
North British Locomotive Company, 120, 154, 160
North West Engineering Trades Employers Federation, 32

Ovenstone, John Keir, 30

Parkes, J. D., 86, 101, 114, 182
Parkhead Forge, 50
Pate, George, 56, 86, 89, 103, 114, 116, 129, 130, 134–138, 146, 147
Pate, William, 86, 100, 114, 148, 149, 162
Paterson, Alexander Nisbet, 40, 41
Penman of Dumfries, 10, 19
Pollock, J., 169, 173, 175, 183

Pope Manufacturing Company, 3, 5
Pullinger, T. C., 43

Rankin Kennedy and Company, 1
Ransome, Ernest L., 140
Reddy, G., 93
Reid, W.C., 146, 152
Richardson, G., 176
Roebuck, J. W., 23
Rough, A., 27

Salter Report, 104, 105, 116, 143
Scottish Motor Traction, 109
Sheffield Depot, 91, 116
Simpson, John, 12
Smith, William Alexander, 44
Spurrier, Henry, 150, 151, 153, 157, 166, 169
Stevenson, S. and Company, 18
Stokes, Donald (Lord), 151, 155, 166, 172, 176, 186
Storry, Alex., 31
Sykes, Meg, 193

Thomson, Robin, 27, 78, 85, 86, 112
Thorpe, H. G., 57
Transport and General Workers Union, 66
Trussed Concrete Steel Company, 40

Verel, W. A., 9

War Office Subsidy Scheme, 33
Watt, James, 108
Webster, T. L., 28, 56, 100
Weir, William, 19, 86
Western S.M.T., 109
Willesden Depot, 73, 129
Wilson, R. W., 27
Wright, Frank Lloyd, 40
Wright7 P. E.7 56
Wright, Dan, 187, 189

Young, G. M., 28, 31, 57, 101, 103, 114, 182
Young's Bus Services, 110–112